To Joyce

with love and hope
of healing

Joan Connell

# HEALING ALL CREATION

## Genesis, the Gospel of Mark, and the Story of the Universe

*Joan Connell and Adam Bartholomew*

ROWMAN & LITTLEFIELD
*Lanham • Boulder • New York • London*

Published by Rowman & Littlefield
An imprint of The Rowman & Littlefield Publishing Group, Inc.
4501 Forbes Boulevard, Suite 200, Lanham, Maryland 20706
www.rowman.com

6 Tinworth Street, London SE11 5AL, United Kingdom



British Library Cataloguing in Publication Information Available

**Library of Congress Cataloging-in-Publication Data**

Names: Connell, Joan, author.
Title: Healing all creation : Genesis, the Gospel of Mark, and the story of the universe / Joan Connell and Adam Bartholomew.
Description: Lanham : Rowman & Littlefield, 2019. | Includes bibliographical references and index.
Identifiers: LCCN 2018052414 (print) | LCCN 2019014309 (ebook) | ISBN 9781538120989 (electronic) | ISBN 9781538120972 (cloth : alk. paper)
Subjects: LCSH: Bible and science. | Creationism. | Healing—Religious aspects—Christianity. | Bible. Genesis—Criticism, interpretation, etc. | Bible. Mark—Criticism, interpretation, etc.
Classification: LCC BS651 (ebook) | LCC BS651 .C6247 2019 (print) | DDC 220.6—dc23
LC record available at https://lccn.loc.gov/2018052414

∞™ The paper used in this publication meets the minimum requirements of American National Standard for Information Sciences—Permanence of Paper for Printed Library Materials, ANSI/NISO Z39.48-1992.

Printed in the United States of America

For Linda and Dean, Nora and Jesy

# Contents

# Acknowledgments

No book is written without the partnership of others. New Testament scholar Thomas E. Boomershine has profoundly shaped my life as a scholar, teacher, and pastor. The Rev. Linda Bartholomew has been my muse, a supportive spouse, a charismatic pastor, and an illuminating critic. The students in my New Testament classes at Gonzaga University have been patient and hospitable partners in the ongoing conversation about the Gospel of Mark and the challenges of the modern world. Gonzaga Department of Religious Studies faculty colleague Kathy Finley has added insights about cosmological theology and the contributions of Catholic writer Judy Cannato and theologian Ilia Delio. The Rev. Rick Matters has directed me to the writings of the seventh-century theologian Maximus the Confessor, modern Orthodox theologian David Bentley Hart, and philosopher William Desmond. Dean Wright has been a skillful manuscript editor, ensuring clarity and consistency in the text.

Adam Gilbert Bartholomew

Thanks are due to many people who helped make this book possible. To Dean Wright and Nora Titone for unwavering love, inspiration, and editorial guidance; to Adam Bartholomew for insight, friendship, and wisdom; to Celia Wexler for opening doors I didn't know existed; to the Daughters of the Heart of Mary for educating me in religion, ethics, and moral issues; to Judith Neuman Beck and the late Deborah Howell for creating the journalistic environments that allowed me to pursue these interests; to Susannah Heschel and Mary Evelyn Tucker for their scholarship and intellectual generosity; to Ira Rifkin and Carolyn Dale for expert editorial advice. And special thanks to Rolf Janke of Rowman & Littlefield for believing in this project and to Courtney Packard and Rossinna Ippolito for their enthusiasm and support.

Joan Connell

# Introduction

From the beginning of human history, storytelling has been central to our understanding of ourselves and our place in the universe.

In every culture, the myths that tell the story of our origins give insights into who we are and why we exist. They shed light on the present and make sense of the dilemmas we face. Stories enforce our values, celebrate our victories, lift up what we hold precious. They illuminate the dark corners of our existence: hurts and harms, sorrows and fears, defeats and despair. They explain why we matter; they help us discern our collective destiny in a continually evolving, ever-expanding universe. Stories hold the possibility of bringing us closer to God as the origin, energizer, and fulfiller of life.

The earliest narratives in the Judeo-Christian tradition were, in fact, simply spoken: dramatically performed by storytellers whose names are lost in time. These oral traditions were passed from person to person and ultimately written down for other storytellers to share. Over the centuries, they have been gathered up into manuscripts, vetted and edited into official publications whose contents varied according to the prevailing language, theology, and political culture. The most ancient stories of the origins of the material world are written in the soils and sand, rocks and waters of our beleaguered Earth. The stories of our cosmic future can be read in the orbits of the planets and in the movements of the stars.

In this book, a scripture scholar and a religion journalist explore the literary, theological, historical, and spiritual symmetries that exist between the creation myths of Genesis and the healing ministry of Jesus as told in the Gospel of Mark, in the context of a much larger story: the evolution of the Earth and its place in the unfolding universe. We look at these stories in the context of cosmological theology, which is our term for a conversation that has been ongoing for centuries by scientists, religion scholars, and theologians. This book focuses on the Judeo-Christian tradition, but cosmological theology (variously known as "the new cosmology," "eco-spirituality," "environmental theology," or simply "theology") is a wide-ranging conversation emerging among a rich diversity of faith traditions: Asian religions, indigenous beliefs, Islam, and the ever-growing cohort of people detached from institutional faith who consider themselves spiritual but not religious. In our definition, cosmological theology expands the frame of inquiry from the merely personal or denominational faith perspective to include our moral and spiritual obligations as the dominant species on an endangered planet: Do we, as a species, have the capacity to evolve away from the evil, greed, animosity, and violence that have estranged us from nature and each other? Why is it so difficult for humans to recognize the sanctity of the material world? How best can we share our planetary home with the other forms of life that dwell here? What role will our species play as we hurtle into a future darkened by our abuse of the Earth? Are we capable of healing the damage? At this time of global moral, political, and environmental crisis, will our destiny as a species be life-affirming or life-impairing? Will we be the agents of the Earth's extinction or of its ability to flourish?

Much of this conversation has unfolded in the rarefied realms and language of academic discourse. It is our goal to bring the ideas and principles of cosmological theology to a general audience—particularly millennials—in whose hands our future rests.

## GENESIS, JESUS, AND THE STORY OF THE UNIVERSE

In the Judeo-Christian tradition, the first depiction of human-wrought harm to the natural world is found in Genesis, when Adam and Eve's disobedience earns a curse that puts them out of harmony with nature and their creator. This plot development symbolizes the source of all the discord and brokenness that permeate the world.

In response to this brokenness, Genesis tells us that God forms Israel to bring healing to the world. (In this book, when we speak of Israel, we speak not of the modern political state, but of the ancient nation that traced its history back to Abraham and Sarah, and who understood themselves to be the people of the God who created the universe.) Israel's self-understanding was nourished over time by a shorter Hebrew and longer Greek collection of scriptures sacred to Israelites throughout the Roman Empire.[1]

In the time of Jesus these people called themselves Israel; outsiders called them Judeans. Both Judaism and Christianity developed out of Israelite sects that differentiated from each other over the course of four to seven centuries.[2] In Genesis and the books that follow it, the very existence of Israel represents its belief in the Creator's determination to restore to the world its original harmony. This would be achieved through the contours of a partnership with the divine described in the Torah.

In our reading of Mark's gospel, Jesus saw it to be his mission to complete God's work by restoring humans to full partnership and returning the world to the conditions of Paradise. In Mark, the message and ministry of Jesus lifts the curse of Genesis and opens a path to healing our collective brokenness.[3] This is likely to be a new insight for many Christians. This is what makes Mark's story of Jesus relevant to the global issues in which we find ourselves anxiously absorbed today. One of the key purposes of this book is to show how, in the course of Mark's narrative, Jesus's mission initiates a quantum leap in Israel's mission of restoring humankind and all creation to its initial state of harmony.

The Gospel of Mark and the Book of Genesis can both be understood within the larger context of the unfolding story of evolution. Generations of scientists, philosophers, and theologians have been telling their own creation story of the Earth and its place in our cosmic home. Modern physics, geology, and biology all yield a story of evolution. In the story of the cosmos the human species has emerged as an almost insignificant latecomer. Insignificant, that is, were it not for the fact that the impact of our species on the Earth has been out of all proportion to the infinitesimal span of our existence.

In partnership with Biblical scholars, some Christian theologians have been getting a clearer picture of Israel's unique contribution to this discussion. The story that emerges in the traditions of this tiny and powerless nation, especially the collection of narratives contained in the Bible, is cosmic in scope. This has led an increasing number of scholars to bring Israel's story into dialogue with the scientific story of the cosmos.

## AN EXPANDING CONVERSATION

Evolution is a foundational narrative for cosmological theologians; they reject the idea that there was an actual time in the past when humans lived in harmony with each other and with nature. The harmonious interplay between humans and nature in Paradise portrayed in Genesis can be seen as the goal toward which the universe is evolving.[4]

Cosmological theology rests in large part in the work of Pierre Teilhard de Chardin (1881–1955), the French Jesuit priest, scientist, and mystic. At first rejected by the church, Teilhard de Chardin's views of Christianity, evolution, and the transformative potential of science and technology were suppressed as heretical during his lifetime, but today have been embraced by the Catholic Church and other faith groups. German Jesuit Karl Rahner did not study Teilhard de Chardin's works, but shared his cosmological perspective.[5]

It was Thomas Berry (1914–2009), an American priest of the Passionist order and scholar with a particular interest in Asian religions, who took Teilhard de Chardin's ideas to the next level, expanding and deepening the understanding that there is a spiritual dimension to evolution and telling a new story about creation, the cosmos, and the sanctity of the Earth. At Fordham University and later at the Riverdale Center of Religious Research, Berry mentored scholars and articulated a compelling vision of environmental activism based on the story of the unfolding universe. Among them were John Grim and Mary Evelyn Tucker of Yale University, who today play a leading role in spreading awareness of Berry's work and that of Teilhard de Chardin. Georgetown University theologian John F. Haught has developed an understanding of evolution that accommodates both science and religion; Franciscan sister Ilia Delio, a theologian who was trained as a scientist, is a dynamic evangelist of cosmological theology, who writes prolifically and lectures widely on the synthesis of science and spirituality. The German Reformed theologian Juergen Moltmann adds a Protestant voice to this growing conversation. British theologian Ursula King, known for her work on Asian religion, mysticism, and gender issues, has written extensively on Teilhard de Chardin, placing his work in the context of Eastern religions.

The late Judy Cannato, a lay Catholic author and spiritual director, was not a trained theologian. Her best-selling books have had a significant impact, particularly on millennials interested in learning more about the relationship of spirituality and science. Cannato's 2010 book, *Field of Compassion*, has the virtue of bringing together in a comprehensive way many of the insights found in the works of those who are more prominent. She has done so in a way that makes them accessible to interested beginners and relates them explicitly to the ministry of Jesus and developments in the theology of the Church.

She sees Jesus as one whose life, teaching, and healing works all had a transformative impact, generating "fields of compassion" that through the actions of his followers widen over time and space. She suggests that the cumulative impact of individual acts of compassion helps heal a wounded world.[6]

The ideas explored in this book reflect only a limited view of the big ideas in the emerging field of cosmological theology. We are indebted to those who have provided an evolutionary lens through which we understand the creation myths of Genesis and Mark's gospel. Viewing these stories of creation, estrangement, healing, and redemption from an evolutionary perspective has led us to revisit some of Christianity's historic failures, from the early abandonment of nonviolence to the institutional church's reluctance, until recently, to embrace and act upon scientific discoveries, particularly as they relate to the environment. We have examined the impact of dispensational theology and apocalyptic thinking, whose vision of the world ending in fiery apocalypse has produced generations of Christians so focused on end times and their personal relationship with Jesus that they have little compassion for others and scant regard for the sanctity of the Earth.

Our reading of scripture, informed by cosmological theology's expansive view of the mysteries of the universe, also leads us to challenge some of the default understandings of Judeo-Christian scripture that prevail today across many traditions: The curse in Genesis that renders women inferior to men is a storyline that continues to perpetuate sexism and minimize the roles women are allowed to play in institutional religion and in society. The emphasis on the dramatic events of the crucifixion in the Christian story often serves to eclipse the importance of the healing ministry of Jesus—and the vital role his followers play as full partners in his mission. The default understanding among many Christians that their relationship with God will only culminate in a distant afterlife blinds them to their duty to—and their delight in—the flawed and fragile Paradise of the present.

It has been said that the Bible is the most purchased and least read book in America. It is our hope that by inviting a general readership to join this unfolding conversation, we will all be able to share in the riches of contemporary scripture scholarship, the wisdom of cosmological theology, and a renewed awareness of the sanctity of all creation.

# 1

# Lifting the Curse

A myth is not a verifiable history. It can be truthful without being factual, a story that articulates a culture's understanding of its basic realities. The first eleven chapters of the Book of Genesis contain the myths the people of Israel crafted to explain the origins of the universe and their own destiny as the chosen people of God.

The Genesis myth begins with the story of the creation of the world, followed by an account of the first human couple.[1] These stories of the moral choices Adam and Eve face and the consequent curses God imposes on them set the stage, centuries later, for crucial narratives in the Gospel of Mark.

The Genesis narrator tells us that God created the first man from the dust of the Earth and the first woman from one of his ribs. The idea that Eve was made from Adam's body, combined with details from her temptation by the serpent, has been used throughout history to support the subordination of women to men. (Understanding how the Genesis story explains the impulse in human history to render women inferior and how Jesus addresses this in the Gospel of Mark carries some radical implications for male-dominated societies, past and present.)

God sets the first couple in a garden filled with fruit trees. Peace prevails and they live in harmony with nature. Life is everlasting; death is nowhere in sight. Adam is not just another of the many creatures God created. He is put in charge of tending the land. He names the beasts of the field and the birds of the air. The human caretakers of this paradise are told that they are free to eat the fruit from any tree. Only one, at the center of the garden, is off-limits: the tree of the knowledge of good and evil. God warns Adam and Eve that if they consume fruit from that tree, they will die. And God builds into these two humans the perplexing freedom simply to disobey the established rules.

To understand what the character of God is doing in this story, consider a contemporary analogy. Every digital device upon which we depend today functions according to an operating system. We curious humans are continually discovering new and even radically conflicting features of our world's OS. In the past century, for example, quantum physics has revealed a whole new subatomic level of reality that operates according to quite surprising rules. In a departure from the Newtonian scientific tradition that saw the universe as a stable and unchanging operation, under the rules of subatomic physics, reality opens onto an ever-increasing array of new and unpredictable possibilities.

In Genesis, God functions as both the creator of the universe and the administrator of its operating system. In the instructions given to Adam and Eve about what is off-limits to them in the center of Paradise, God is naming a key part of the OS that humans must not tamper with if the world is to function smoothly. But the first couple chooses not to honor that limit. Their choice introduces a whole new turn of the cosmic plot and raises some intriguing questions: What are the consequences for this failure to go along with the program? Are the consequences irreversible? Is there any larger storyline into which their disobedience introduced a disruption? These are the questions that occupied the Christian heirs of Israel's scriptures. They struggled to understand the relationship between the story of Jesus and this larger narrative. And over the centuries, different kinds of Christians have answered that question in different ways.

## CHOICE AND CONSEQUENCE IN PARADISE

In her 2008 book, *Christ in Evolution*,[2] cosmological theologian Ilia Delio writes that from the time of St. Augustine of Hippo in the fourth century, most Christians in the West had narrowed their attention to Jesus's work of healing the rupture between humans and their God. This healing is accomplished by the birth of Jesus, whose mission on Earth is solely to atone for human sin. In Western Christianity, the significance of the Genesis story itself was limited to providing the account of the Fall into sin and the human need for redemption. In the default version of Christianity, Adam's disobedience was perceived as a necessary stumble, without which Jesus would never have walked the Earth.[3]

The myths of Genesis are told in a style that provokes questions in the imagination of the reflective listener. Is Adam and Eve's disobedience really the will of God? Is this how God intended the plot to unfold in the first place? Was there perhaps a different original goal that got derailed? Did God create the world to be a placid paradise in which the humans would simply have the job of keeping things operating smoothly in an uninterrupted equilibrium? Is the only problem with Adam and Eve's disobedience their broken relationship with God?

The Genesis narrator makes it clear that the first couple's disobedience has repercussions throughout the entire garden of Paradise.[4] The vast network of harmonious relationships in the cosmos begins to malfunction. All those relationships need to be repaired. But then what? Would repair simply restore an original equilibrium? Or was Paradise brimming with unimagined potential when God placed the couple there? Does God create Adam and Eve not just to maintain a finished garden but to be partners in developing its potential?

A close reading of the choices the first humans face in the Genesis myth reveals something quite different from the way Christians traditionally interpret it. In Genesis 1, God declares creation to be good. In Genesis 2, God names the possibility for evil alongside the good. So it is actually God who introduces disequilibrium into the narrative by commanding the couple not to eat from the tree of the knowledge of good and evil. It is God who raises the question of the future and who introduces two ways it can unfold: the split paths of obedience or disobedience.[5]

Adam and Eve's rebellious choice results in a narrative familiar to all human beings. Their newfound knowledge of good and evil is not a mere theoretical possibility: their knowledge becomes their experience.[6] To know evil is to experience the way things can go wrong. In place of harmony there is violence. The satisfaction of desire leads to more desire. Getting enough to eat today means you will be hungry again tomorrow. Life is a constant struggle and it finally ends in death.

But what if Adam and Eve had not made that choice? Would life in the physical cosmos be one of humdrum repetition, of maintaining a finished, perfect system? If God created the material cosmos as more than a backdrop for the story of the divine-human relationship, does God have more in mind for its future than mere maintenance?

If Adam and Eve had not asserted their independence and explored evil as well as good, would the story of the world lack the creative tension that makes a compelling narrative? Under that scenario, humans might be nothing more than well-programmed robots in a paradise where nothing ever changes, unfolding in strict conformity to a predetermined plan.

In *The Beauty of the Infinite*, Eastern Orthodox theologian David Bentley Hart proposes a whole alternative plot for the ongoing story of life in Paradise: the unfolding of beauty in infinitely new and fascinating manifestations. That such potential is built into creation is clear from God's command, "Be fertile and multiply" (1:28). In Hart's view, beauty evokes in the beholder an energizing desire. Desire revels in difference. Desire involves dramatic tension: the ongoing challenge of the restraint required to preserve the differences that are essential to ongoing fascination. The beautiful beckons and invites. And in fact, what is beautiful may not be immediately appealing on the surface. One may need to develop a taste for it. But once the beautiful has captured our interest, we may find that we are unable to let it go until it has blessed us. Encountering beauty, Hart writes, does not leave us as we were. Our whole self is changed.[7]

In contemporary thought, people often imagine the self to be a finished identity that is uniquely ours, but hidden from us through social conditioning and in need of being rediscovered. For Hart, it is in the presence of beauty that the self is developed. He writes that the "'self' is that 'matter' in which beauty impresses itself, that 'place' where the light of the other, and of all being, shines, gathered in a reflective surface of incalculably various sensitivity (the physical senses, thought, imagination, anticipation, memory, desire, fear)."[8]

The seventh-century Eastern Orthodox theologian Maximus the Confessor offers another view of the choices inherent in Paradise, in a work entitled *Ambiguum*.[9] In this view, Adam and Eve's desire for the knowledge of good and evil stems from the mistaken belief that they could choose God properly and freely only if they also understand the alternative, namely evil. But Maximus believed that one must choose the beautiful simply because it is beautiful in itself, not because it is more desirable than the alternative. The beauty of another beckons the beholder into an unfathomable mystery, in which every discovery opens onto yet another mystery.

Fascination with the unfathomable beauty of the universe is in fact a fair description of the adventure of scientific discovery and stands as yet another view of Paradise. Science is devoid of the plotlines of violence and conflict we expect if a story is to capture our interest. Yet it is in no way lacking in excitement. In contrast to the rapacious employment of technology in which the universe has been cast in the role of an opponent to be conquered, enslaved, and ravished, the story of modern science has unfolded along a completely different trajectory: a sense of wonder as the beauty of the universe unveils itself before the disciplined gaze of the scientist. The scientist is in fact a lover who is patient and imaginative and restrained and respectful enough not to intrude upon beauty's allure but only to stand in awe. Hart compares the ongoing differentiation and complex new combinations of the evolving universe to music that modulates into ever newer harmonies. The tiny miracles of Johann Sebastian Bach's *Goldberg Variations* can be seen as a microcosm of the music of the spheres.

## THE MAJESTY OF CREATION

Theologians down through the ages who have drawn inspiration from Francis of Assisi's veneration of nature share Hart's sense of awe at the beauty of Earth and the majesty of the surrounding cosmos. This understanding of the sacredness of all creation is quite different from a competing narrative in Western thought that reduces the material world to merely an arena in which humans struggle to define their relationship with the divine.

Those who resonate with Francis's veneration of nature intuitively see the role of Christ to be far more than restoring disobedient human beings to an unbroken relationship with God. And Eastern Orthodox theologians are more inclined to see the conflict Adam and Eve perpetrated in the garden as a key to a story larger than simply their disobedience. Maximus the Confessor was one of them. He recognized that the visionaries and prophets of Hebrew scripture hoped for a renewal of all creation;[10] so did the Israelites who came to believe that Jesus had brought their visions and prophecies to fulfillment. But the default narrative of Western theologians down through the centuries reflects the Platonist view that focused its attention on the afterlife. The central reality of Christianity for those theologians had little to do with the natural world. Their concern was the redemption of the human individual: the suffering, death, and resurrection of Jesus would result in the forgiveness of human sin that would admit the disembodied human soul to eternal life in a far-distant spiritual heaven.[11]

Contemporary theologians like Pierre Teilhard de Chardin also envisioned a cosmic reconciliation through Christ to be the goal of the story of creation. The contemporary German Reformed theologian Juergen Moltmann describes the universe as an open system that evolves not out of past events in a unilinear chain of cause and effect, but in a widening web of elementary particles and increasingly complex structures yielding an ever-growing range of possibilities.[12] Moltmann derives his description from quantum science. The Italian physicist Carlo Rovelli describes the workings of the evolving universe in this way. Subatomic particles, he writes,

> do not have a pebble-like reality but are rather the "quanta" of corresponding fields. . . . They are elementary excitations of a moving substratum. . . . Minuscule moving wavelets. They disappear and reappear according to the strange laws of quantum mechanics, where everything that exists is never stable and is nothing but a jump from one interaction to another.[13]

The view of modern scientists is that the world has evolved over billions of years and is still evolving. From the unfolding process of evolution has emerged a human species with a unique mental capacity for seeing how things are and imagining how they can be different. Through our free imagination, our species has added a whole new layer to the fruits of the evolution of the Earth. We have covered large areas with constructions of our own devising that far exceed our requirements for survival. We have clothed our bodies in an endless variety of ways and developed forms of social organization far more complex than other creatures.

The truths embedded in the Genesis myth resonate with our contemporary understanding of the role humans play in the great story of the evolving universe. Along with the freedom to contribute to the creative unfolding of the universe also comes the capacity to disrupt the cosmic operating system and do enormous damage.

Over the centuries, humans have operated under the collective delusion that there are no limits to our power and no consequences to our actions. The natural world has been regarded as a resource to mine, manipulate, and abuse.

And now we are plunged into crisis: we must decide as a species whether we will continue our role as agents of the Earth's evolution or end it by destroying the environment and rendering extinct the living creatures who call it home, including ourselves.

The narrator of Genesis recognized that we are a species uniquely endowed with both the responsibility for creation and the potential to create chaos, brokenness, and death. God's command to Adam and Eve is not an arbitrary manifestation of a desire to be in control. It is an invitation to humanity to be caretakers of the Earth and all its living creatures, with a clear freedom to develop it beyond what it was when God introduced us into the scene. In the exercise of our freedom we must recognize and honor the limits imposed by the natural world itself.

In the myths of Genesis and in the story of the universe told by modern science, the universe is not a closed system of impersonal cause and effect, in which humans are puppets whose sense of freedom is an illusion determined by the chemical and electrical processes in the brain. Each story can be read as that of an open system in which the human creature has the capacity to be creative or destructive. Israel imagined the creator of this system to be a sovereign who desired a partner free of control, with whom to share the great adventure of the evolving universe. This freedom involved a great risk. The story of the Bible clearly recounts the consequences of that risk—consequences Western civilization has ignored for centuries but which are now clearly real and threatening to our continuing role in the evolution of the universe.

In the Genesis story, faced with the choice between creative partnership and going their own way, the first humans chose the latter. The rest of the Christian Bible is the story of the Creator's response to that choice. And there is no guarantee in science or the Bible that the human species will turn out to be the responsible partner for whom the narrator of Genesis hoped.

## GOD'S CURSE

Adam and Eve's decision to violate God's command plays out in a drama of temptation, with a serpent in a starring role, urging a curious couple to taste the fruit of a particular tree. The consequences of that decision were swift and severe, delivered in the form of a curse from God, first to the serpent and then to the offenders themselves. It is the Genesis storyteller's explanation of the origin of every human misery: fear and subjugation, the pain of childbirth, loneliness and sorrow, war and disaster, suffering and death.[14]

In Genesis 3:14–19, after Eve is persuaded to convince Adam that they should taste the forbidden fruit, God says to the serpent:

> Because you have done this,
> cursed are you
> among all the animals, tame or wild;
> On your belly you shall crawl,
> and dust you shall eat
> all the days of your life.
> I will put enmity between you and the woman,
> and between your offspring and hers;
> They will strike at your head,
> while you strike at their heel.

To the woman he says:

> I will intensify your toil in childbearing;
> in pain you shall bring forth children.
> Yet your urge shall be for your husband,
> and he shall rule over you.

To the man he says:

> Because you listened to your wife and ate from the tree about which I commanded you, You shall not eat from it,
>
> Cursed is the ground because of you!
> In toil you shall eat its yield
> all the days of your life.
> Thorns and thistles it shall bear for you,
> and you shall eat the grass of the field.

By the sweat of your brow
  you shall eat bread,
Until you return to the ground,
  from which you were taken;
For you are dust,
  and to dust you shall return.

God's curse enumerates the broad scope of conflicts that permeate human existence. Notice the conflict between humans and animals in the image of the snake that bites the heel of the human and the human who strikes the head of the snake. Contrast this to the harmony that existed among all living creatures before Adam and Eve succumbed to the serpent's temptation (Genesis 2:18–20).[15]

The Genesis storyteller leaves us free to imagine that snakes were not originally created to slither along on their bellies. He implies that originally they moved along in some other way. In images of the temptation of Adam and Eve, the snake is sometimes pictured as walking on legs. (A good example is *The Temptation*, painted in 1470 by Hugo van der Goes.) The point is that the snake now moves along the ground because something has gone seriously wrong with the created order.

The curse upon Eve for her disobedience is similar. Pain in childbirth demonstrates a conflict between humans and nature. Why is pain in childbirth a curse and not just a part of human life? Is the Genesis storyteller observing a difference between humans and animals and concluding that something had gone wrong as a result of the humans' decision to say no to God? Does the storyteller long for a time when childbirth would be as effortless and natural for humans as the birth process is for other animals?

Entwined with this painful anomaly in the natural order is the irony that despite the pain of childbirth, a woman's desire will be for her mate. This certainly is true to human life. To experience pain as a result of the actions of the one you love and yet to continue to desire that person is manifested in more than childbirth. It is an unavoidable dimension of any close human relationship—lovers, parents, children, siblings, or friends. Its extreme manifestation is in abusive relationships. Whether it is an abused child or an abused spouse, the victim typically has a difficult time giving up hope that things will change.

Adam is cursed to a lifetime of hard labor, drawing sustenance from a sullied Earth. The fertile soil of Eden, which produced food in abundance, is no longer available. Adam finds himself in a hard, parched landscape, dense with stones, thorns and thistles.

The final aspect of the curse is death itself. Life was everlasting until Adam and Eve violated God's rule. But with this curse, life becomes finite; humans would return to the ground from which they were taken.

How does God pronounce this curse? With what feeling and tone of voice? Is God angry or grieved when uttering these words? Perhaps a bit of both. There are no clear descriptions in Genesis for how the oral storyteller made God sound. But one thing is clear: this despoiled Earth, full of conflict, pain, and broken relationships, is not what God had in mind for the world or his creatures. The narrator of Genesis is not happy with life after Paradise. One of the more striking instances of this is male domination of women, which over time in many cultures has become widely viewed as natural and proper. For the Genesis storyteller, this domination is neither natural nor proper. It could be said that the inequality of man and woman is a curse that even the author of Genesis yearns to be lifted.

## HEALING THE WORLD

As the story unfolds in Hebrew scripture, God takes steps to re-create a humanity that will restore the cosmic order and heal the world. First he selects one righteous man, named Noah, along with his family, and pairs of animals and birds, who are saved from a great flood that has wiped out all remaining life, human and non-human. That does not work. The moment the flood recedes, Noah and his family return to the same destructive behaviors.[16]

Then God calls Abraham and Sarah out of Ur of the Chaldees to be the progenitors of a new nation, Israel; God reveals his plan in the form of the Torah given through Moses. That yields spotty success, a situation recognized first by the Hebrew prophets, and later not just by Christian Israelites but by a number of other Israelite sects in the time of Jesus. Because God's people were then living under the domination of the Romans instead of in freedom with God alone as their king, the Pharisees, Zealots, and Essenes all believed that Israel had not yet conformed their lives according to the Torah. It was because of that failure that God had their ancestors sent into exile. Historically, that exile began in 587 BCE in the foreign land of Babylon, but it had become a domestic exile as well. Half a century after the deportation to Babylon, the Persian conqueror of Babylon permitted the deportees to return to their ancestral land; but for most of the next six centuries they continued to live under foreign governments. Rome was only the latest. Attributing their ongoing exile to their ongoing sin, each sect had its own visions for how to bring the life of Israel back into harmony with God.[17]

The storyteller we call Mark was a member of the sect of Israelites that believed Jesus of Nazareth was God's Messiah. The organizing principle of Mark's gospel is its account of how Jesus sets about the task of healing of the world. According to Mark's story, Jesus comes onto the earthly scene as a human being in complete harmony with God's will, a harmony that the Pharisees also sought through the perfect and joyful fulfillment of the law. Unlike Adam and Eve, Jesus manages to resist Satan's temptations to disobey the will of God. Soon after, during Jesus's riverside encounter with John the Baptist, a voice is heard from heaven: God declares Jesus to be "my beloved Son." Before Jesus embarks on his ministry of healing and reconciliation, he declares that this harmony between himself and the Creator is about to be restored throughout the whole earth (Mark 1:14–15). The longstanding curse that set humans apart from God begins to lift.[18]

At various points in Mark's narrative, we see dramatic and auspicious instances of Jesus's healing efforts to lift the curse of Genesis. Mark's story of Satan's temptation of Jesus in the wilderness is clearly connected to the serpent's temptation of Adam and Eve. Unlike the gospels of Matthew and Luke, Mark is silent about the details of Satan's offers. But by refusing Satan's unknown enticements, Jesus reintroduces into the world the impulse to act in harmony with God's will. If we keep that in mind, Jesus's acts of healing the sick, feeding the hungry, and reconciling the alienated can be seen as remedies to the ills of a world that has been cursed.

The hard labor on thorn-infested ground to which God cursed Adam describes well the life of Jesus's first-century listeners. Jesus's seemingly effortless acts of feeding enormous crowds of people, with as much bread and fish as they want, signals the reintroduction into the world of effortless sustenance in the garden before the Fall.

CHAPTER 1

## FIELDS OF COMPASSION

In her 2010 book, *Field of Compassion*, Judy Cannato likens the impact of Jesus's acts of ministry on the human and natural realms to a "morphogenic field of compassion" that has the power to heal and transform the world. The term originates in cellular biology: it is the ability of an organism to shape the next generation by learned behaviors that result in non-genetic changes to fields of information and memory. These are established through interaction of the organism with the environment, imitated by other members of a species, and open to being eventually supported by random genetic mutation.[19]

In science, a morphogenic field starts out small, with one or a few individuals acting in a novel way. As others of the species imitate this new behavior, it spreads more and more widely through the species. Jesus's work started out small, but history shows that it spread widely and rapidly. The ongoing efforts to advance human rights, end slavery, and liberate women from male domination, and the emerging awareness that we must restore a harmonious relationship with nature, all hint at a transformation rooted in the teachings of Jesus.

In *The Unbearable Wholeness of Being*, Ilio Delio describes recent developments in science regarding cooperative activity among members of a species that fit well with Cannato's spiritual spin on biological phenomena.[20] Newtonian science envisioned reality as a finished system of independent entities interacting according to fixed laws. But quantum physics envisions reality as something more like fields of energy and interaction. Similarly, on the biological level, individual entities appear to be not genetically finished, closed, and independent but open systems responsive to the larger systems that surround them in a kind of improvisational dance. Given the role of novelty in evolutionary development, modern thinkers since Teilhard de Chardin and Julian Huxley before him are more open to considering the presence of some degree of consciousness and intention in the process, perhaps reasoning back from the "quantum leap in consciousness" exhibited in the human species.[21]

The transformative actions of Jesus in the Gospel of Mark are compassionate, healing, and path-breaking.[22] He is both an evolutionary and a revolutionary force. He feeds the hungry, heals the sick, reconciles the alienated, and identifies with the victims of violence; he casts out demons and embraces his enemies.

He counsels sinners to repent their actions and trust the dawning of God's rule. His most radical innovation is the commitment to refrain from violence against enemies even at the expense of his own life and the success of his mission. These acts generate the "fields of compassion" that have the power to alter human history and change the shape of the world.

The transformation is far from complete; no one can look at the world today and say with confidence that the curse that estranged us from all creation has been lifted. Our liberation from the curse of Genesis occurs in fits and starts, plays out in our individual psyches and our collective history, in cycles of progress and setback.

During the first three centuries of the Church's history, Christians self-consciously embraced Jesus's rejection of violence when faced with persecution and the threat of being killed for their faith. But since the fourth century, when the Roman Emperor Constantine embraced Christianity and made the Church a partner in administering the Empire, the rejection of violence, the absolute center of Jesus's strategy for lifting the curse, has all but disappeared from the awareness of most Christians.

There is a rising sense of urgency among theologians about the need for a spiritual transformation. In their 2015 book, *Seven Revolutions: How Christianity Changed the World and Can Change It Again*, Michael Aquilina and James L. Papandrea warn that if we do not recognize and embrace the world-changing principles of Christianity, we risk a return to the beastly conditions of pre-Christian Roman society.[23]

That it would be Jesus's mission to generate fields of compassion to heal this world is a revolutionary notion for Christians in the West. For almost our entire history we have limited our attention to the drama of Jesus's death on the cross in order to secure our redemption. Jesus's concrete acts of exorcism and physical healing and feeding and reconciliation of enemies have been understood simply as manifestations of his divinity and his love. Their relationship to Israel's story of the creation of the world in Genesis has been completely missed. As our world becomes increasingly unstable and governments respond to emerging crises with shows of force, Jesus's concrete efforts to reconcile enemies and his ultimate refusal to employ violence as a strategy for achieving victory are rarely considered.

Jesus's healing, nurturing, and comforting actions in Mark's gospel may seem small compared to the enormity of the world's problems. Perhaps it simply never occurred to Christians that Jesus could be seeking to do anything so grand in scope as to transform a broken world. But by trusting Jesus and receiving his Spirit, people of faith have created fields of compassion that have the potential to transform the world: they have fed the hungry, set up hospitals, reformed prisons, opposed slavery, taught people how to raise their own food, educated the young, encouraged generosity, protested war, defended people on the social margins, and more.

It is becoming increasingly clear that it's not only other humans who require the Church's ministry of healing from the wounds of human violence; the Earth and its non-human creatures are also in need of healing. The violence we perpetrate upon nature goes hand in hand with the violence we perpetrate on other human beings. The interlocking nature of these two realms of human violence has been made dramatically clear in the recent encyclical of Pope Francis, *Laudato Si': On Care for Our Common Home.*

Jesus's story, as told in the Gospel of Mark, also clearly aligns with the story of the unfolding of the universe being told by modern science. As science has developed it becomes increasingly clear that everything in the world is connected.[24] Physically and spiritually, small actions can affect the entire cosmic operating system. If it is possible for a solitary person to generate great evil, it is equally possible for a solitary person to initiate enormous good.

Science echoes the validity of this view. In 1961, mathematician and meteorologist Edward Lorenz discovered what he called the "butterfly effect."[25] In making calculations to predict the weather, Lorenz found that if he varied his calculations by only a few more decimal points, a difference as slight as the fluttering of the wings of a butterfly, the effect on the forecast was enormous. In a similar way, Jesus's small acts of healing and exorcizing and feeding and reconciling a relatively few people generated great changes in the social system. Jesus's deeds inspired those who encountered him and those who later heard the stories about him with the desire, creativity, and courage to take up his ministry in their own lives. The result of this is that the world is different. It is being transformed.

Jesus's ministry of healing, nurturing, and reconciliation is not a mere prelude to his redemptive mission of dying on the cross for people's sins. It is core to his mission. His suffering and death were the most dramatic of a series of compassionate actions meant to lift the curse of Genesis, heal the human family, and transform this world into the Kingdom of God. These actions restored to the story of our origins the human-divine partnership that the Creator in Genesis had originally envisioned: a shared engagement in the unfolding of infinite beauty.

# 2

# The Gospel of Mark

The author of the Gospel of Mark is less of a person than a disembodied voice, an anonymous storyteller who, in the decades after Jesus died, recounted in oral narratives the public life of a man he believed to be the Messiah, capturing the attention of people willing to stop and listen. His story came to be treasured by a broad swath of early listeners who were persuaded to put their trust in Jesus and become his followers. A written text of Mark's story came to be included in the Church's collection of Christian works called the New Testament. In this way Mark's narrative has been passed down through the centuries. Over time, Christians have read and heard it as a source of illumination and inspiration.

Why read or listen to Mark's story today? What can this ancient story mean for us? The answers emerge from the work of an ecumenical community of Biblical scholars and Christian theologians that reaches back to the earliest days of the church, but entered a creative new phase about three hundred years ago.

In the eighteenth century, biblical scholars in Western Europe and America began subjecting the Church's traditions to critical examination. The view of Mark's gospel that has emerged over time resonates with recent efforts of theologians to bring the Christian theological tradition into conversation with modern science in order to address the vital issues of the twenty-first century.

The story of Jesus as it unfolds in the study of Mark's gospel significantly differs from the default telling of the story of Jesus in the Apostles' Creed, the Nicene Creed, and the traditional Eucharistic prayers, hymns, and anthems with which many Christians of the Western Church recite weekly the story of their faith.

The great summaries of the Christian story with which believers in the West are most familiar barely mention God's work of creation. They largely ignore the narrative in the Hebrew scriptures of God's relationship with the chosen people of Israel. The story of Jesus tends to move directly from his birth to his torture and death. These liturgical summaries skip over the account of Jesus's earthly ministry. They understand his death to be an action designed to admit us to eternal life in another world, a non-material heaven where disembodied souls can be at home when believers leave their earthly bodies.[1]

In Christianity's traditional story, the healing of all creation does not show up as Jesus's goal. In fact, because Western theology largely ignores the fate of the Earth, modern Western Christians have historically looked upon the natural world as a realm for humans to plunder, since they will ultimately leave it behind upon death for a greater reward in heaven.[2]

This default interpretation of the life and death of Jesus has its roots in the Greek culture and philosophy that shaped the imaginations of many Gentile converts and told a story of the cosmos that was different from the one told by Israelites like Mark. Under the influence of the Athenian philosopher Plato and his successors, such as the third-century Neoplatonic philosopher Plotinus, many Christians came to believe that the material Earth is not our true home. They believed the human soul originated in an unchanging spiritual world above the Earth. From this realm our souls descended and became imprisoned in our physical bodies. To it our souls would return upon death. The Earth and the human body were not destined for transformation but would be left behind for a different world, already free of corruption because it is non-material and therefore unchanging. The early Gentile converts, brought up with this Platonic story and uninformed of the alternative narrative told by Israel, interpreted isolated passages and terms in the Bible to fit their version of the story.[3]

The history of Christian theology is complex. Greek theologians like Irenaeus of Lyons, who lived in the second century, and Maximus the Confessor, who lived in the seventh century, did recognize the difference between the story told in the Bible and the story told by Platonic philosophy. They included the entire cosmos in their view of Christ's work of redemption.[4] So did later Western theologians such as Alexander of Hales and Bonaventure, both members of the Franciscan order who lived in the thirteenth century and were influenced by Francis of Assisi's sense of kinship with all of nature. These theologians adapted Gentile philosophy to the cosmic perspective of the Bible rather than the other way around. So have recent theologians and liturgical scholars informed by the discoveries of recent scholarship.

But the idea of the cosmic significance of Jesus did not stand a chance against Augustine of Hippo, who flourished in the late fourth and early fifth centuries. Under his powerful influence, theology in the West came to limit its attention to the human need for salvation from its sinful condition. This focus essentially narrowed the story of God's incarnation in Jesus to his death, to save humans from sin and unite their disembodied souls with God in heaven. The created cosmos was nothing more than the stage for this human drama. At least on a popular level, the story that came to dominate the Christian imagination was the Platonic story of the soul leaving the physical body and material world for eternal life in heaven. This narrative still dominates the understanding of the vast majority of Christians.

## MARK'S STORY OF JESUS

Mark told the story of Jesus not as a man who came from heaven to die for our sins so that we could spend eternal life in another world, but as the completion of Israel's story of God's work: bringing healing to a broken creation and calling all people to participate in that work.

Many scholars believe that Mark told his story either during or in the immediate aftermath of the Great Revolt in Palestine against the Romans in 66–70 CE. Inspired by the story of the Exodus from Egypt a millennium before and by the success of the violent Maccabean revolt against the Seleucid Greek rulers two centuries earlier, the entire land of Palestine rose up in violence against the Romans three decades after Jesus.[5] Faithfulness to the will of God required freedom from foreign rule, not ongoing exile in their own land. The stunning failure of this revolt had repercussions for Israelites living in all areas of the Roman Empire. In his commentary on Mark's passion and resurrection, Thomas E. Boomershine convincingly argues that Mark's purpose was to offer Israelites a "Messiah of Peace."[6]

Three decades before the Great Revolt, Jesus offered his people a powerful antidote to the periodic outbreaks of violence that had marked their relationship with Rome since the time of his birth, when Herod the Great was the puppet king of the Romans in Palestine. Jesus showed his fellow Israelites an alternative to violence as the path to faithfulness to their God. In place of division and violence, he sought to reconcile Israelite factions to each other and all of Israel to the Gentiles. Peace in place of ongoing division and violence was the most prominent kind of healing the Earth was yearning for in Jesus's day. It continues to be so in our day. Much of the rest of Jesus's mission brought healing to the damage that war and conquest left in its wake. Hunger, physical sickness, and mental trauma were all exacerbated by the violence divided peoples perpetrated upon each other. Jesus saw his mission as bringing to the threshold of fulfillment God's work of healing the world begun in the creation of the Chosen People of Israel.

For a long time, the almost universal view of scholars has been that Mark's original audience consisted of Gentile Christians. But there is increasing evidence that Mark's first listeners were Israelites who were not yet believers but who were drawn to the message of Jesus because they were seeking new ways to address the problems of their lives by living in greater faithfulness to God's will.

Mark's narrative is saturated with allusions to Israel's traditions that would most likely be recognized by people who had been brought up in the synagogue communities that dotted the Roman Empire. It is also true that Gentiles often attached themselves to those synagogues and learned the traditions as well. Such Gentiles might have been in Mark's intended audience.[7]

Listening to Mark's story without getting his allusions to Israel is like people today watching a play by Shakespeare and not knowing anything about the Bible, Greek and Roman mythology, and history. You can get the story line, but you miss a lot. It is hard to believe Mark designed his story for any but people who would identify with his allusions and understand the details of the story in relation to the larger narrative.

In addition to allusions to Israel's traditions, Mark includes terms that had special meaning to Israelites caught up in the Great Revolt. The Greek word *lestes* is the word for those who were crucified along with Jesus and also in Jesus's quote from Jeremiah (Mark 15:27; Jeremiah 7:11) when he explained his reason for disrupting business in the Temple before his crucifixion.

Is it not written:
"My house shall be called a house of prayer for all peoples"?
But you have made it a den of thieves (*leston*). (Mark 11:17)

This is the term the Israelite historian Josephus uses for the revolutionaries who three decades after Jesus's crucifixion tried unsuccessfully to drive the Romans out of Palestine, taking one of their last stands in the Temple that served as a fortress. The term may well have brought the revolt of the Judaeans to mind for Gentiles. However, for Israelites living outside Palestine and labeled by their Gentile neighbors as "Judaeans," the term must have seemed shockingly prophetic of the Great Revolt and its repercussions for their relationship with Gentiles.[8]

## THE LANGUAGE AND MEDIUM OF MARK'S NARRATIVE

Where did Mark's original audience live? Tradition says first in Rome and then in Egypt.[9] Both places are certainly possible. It is most likely that Mark spoke his story in Greek. Greek was the *lingua franca* of the Roman Empire. Ever since the cream of Judean society had been exiled to Babylonia six hundred years before Jesus, most Israelites, perhaps 90 percent of them, lived outside of Palestine. Few apparently knew Aramaic or Hebrew. All most likely knew the language of the particular area in which they lived. Most likely they also knew Greek, since that language was as common to the Empire as English is common around the world today.

How did Mark get his story of Jesus out to those he wanted to hear it? Here, too, Boomershine has pioneered a new view of Mark's means of communication. The medium in which Mark's story has come down to us through the centuries is the written word. But in Mark's world one did not write a story and then get copies printed to distribute for others to read. Mark's world was one of oral communication; very few people could read. If someone did produce a written text to be read, it was almost always read aloud. Publication meant reading aloud to a listening audience, usually in a dramatic style, not only to hold the listeners' attention but also because the meaning of a story is as much in the feelings generated as in the details of what happened or what was said.

In fact, stories were often composed in the memory of the storyteller without the use of writing. What was written was often a transcription of a spoken narrative that had been composed from memory. When manuscripts were available, they were often memorized in order to be performed. The task of reading from a scroll in public was cumbersome. The text often lacked any sort of punctuation; it required the storyteller at the very least to practice before performing it, just as a musician today practices a score before playing it in public. In both cases one must not only practice the words or notes, but work out the expression.

This view of Mark and his audience is based primarily on evidence internal to the narrative itself, assessed in the context of the cultural and political history of Israel in the first century. An early tradition that can be documented as far back as the second century provides possible confirmation and supplements to this view.

Eusebius, the fourth-century bishop of Caesarea, records in his *Church History* a report of what Papias, the second-century bishop of Hierapolis, learned about Mark from a certain elder of the Church named John:

> Mark, having become the interpreter of Peter, wrote down accurately, though not indeed in order, whatsoever he [Peter] remembered of the things done or said by Christ. For he [Mark] neither heard the Lord nor followed him, but afterward, as I said, he followed Peter, who adapted his teaching to the needs of his hearers, but with no intention of giving a connected account of the Lord's discourses, so that Mark committed no error while he thus wrote some things as he [Peter] remembered them. For he [Mark] was careful of one thing, not to omit any of the things which he had heard, and not to state any of them falsely. (Eusebius, *Church History* 3.39.14–17)[10]

This is a significant text, full of nuance and insights into the practices of the early church. The presbyter John, whom Eusebius quotes, clearly states that Peter "adapted his teaching to the needs of his hearers." He also says that Peter had "no intention of giving a connected account of the Lord's discourses," that is, his teaching. In fact, Mark's story contains only two collections of teaching that we might characterize as discourses, namely the series of parables that Jesus tells in Mark 4 and Jesus's vision of the future in Mark 13.

In an earlier part of his *Church History*, Eusebius gives an account of how Mark's gospel began as an oral composition, which Peter performed for listening audiences who insisted that Mark write it down.

> And so greatly did the splendor of piety illumine the minds of Peter's hearers that they were not satisfied with hearing once only, and were not content with the unwritten teaching of the divine Gospel, but with all sorts of entreaties they besought Mark, a follower of Peter, and the one whose Gospel is extant, that he would leave them a written monument of the doctrine which had been orally communicated to them. Nor did they cease until they had prevailed with the man, and had thus become the occasion of the written Gospel which bears the name of Mark. . . . Clement in the eighth book of his Hypotyposes gives this account, and with him agrees the bishop of Hierapolis named Papias. And Peter makes mention of Mark in his first epistle which they say that he wrote in Rome itself, as is indicated by him, when he calls the city, by a figure, Babylon, as he does in the following words: "The church that is at Babylon, elected together with you, saluteth you; and so doth Marcus my son" (1 Peter 5:13). And they say that this Mark was the first that was sent to Egypt, and that he proclaimed the Gospel which he had written, and first established churches in Alexandria. (Eusebius, *Church History* 2.15.1–2, 2.16.1.)[11]

Note that the first part of this report, which Eusebius found in Clement's Hypotyposes, concurs with the view of many contemporary scholars that the gospel originally was a spoken story, performed by Peter for listening audiences. Some years later at the insistence of Peter's listeners Mark recorded it in somewhat adapted fashion in writing. It is not clear in Clement's report whether Mark performed the story as well or merely put in writing what people had heard from Peter. The report suggests that the reason for writing was to guarantee preservation. Eusebius goes on to say that Papias supports Clement on this. Notice that there is no indication in either of these reports from Eusebius that Peter's listeners were already Christians when they first heard the story.

Although Clement's account does not say that Mark's written text was also orally performed based on what he heard from Peter, Eusebius does go on to call attention to a sentence in the First Letter of Peter, which is included in the New Testament:

> The church that is at Babylon, elected together with you, saluteth you; and so doth Marcus my son. (1 Peter 5:13)

Eusebius explains that Babylon is an epithet for Rome. This is probably because both Rome and Babylon dominated Israel, and by the time of both Eusebius and the author of First Peter both empires had destroyed Jerusalem and the Temple.

Read Eusebius's words carefully: He concludes from this identification of Babylon with Rome that Peter wrote his letter in Rome. This places the man Peter names as Marcus in Rome as well. Eusebius assumes that this Marcus is also the author of the gospel attributed to someone named Mark. He then goes on to say that Mark went from Rome to Alexandria to proclaim the gospel he had written. There is no question here that Mark also did perform the gospel in Egypt. Again there is no indication that Mark's listeners were already Christian. The sequence of events reported by Eusebius is essentially this:

- Peter orally composes and performs a narrative about Jesus in Rome.
- Peter's listeners in Rome insist that Mark write it down.
- Mark adapts what he heard from Peter in a written narrative.
- Mark travels to Alexandria in Egypt and performs the narrative there.

Contemporary conservative scholars and believers are inclined to accept this evidence at face value. Scholars influenced by the critical tradition that began in the mid-seventeenth century and that dominates every academic discipline today raise many questions: How do we know that the author of the gospel titled "according to Mark" was the same as the Marcus named in 1 Peter? Was Eusebius correct in assuming this? Was Peter the disciple of Jesus actually the author of the letter attributed to him?

There is a tendency in the ancient Church to assume that a person of importance who is named in one report is the same person as someone who bears the same name in another report. That may or may not be true. Good scripture scholarship requires the same verifiable evidence and disciplined analysis we normally expect from historians in other fields. Whatever we find in ancient sources, it is a sound principle of scholarship to ask whether uncertain external testimony about an ancient document fits with what we can observe from the evidence internal to the document.

A significant group of skeptical biblical scholars active in the late twentieth century, known as the Jesus Seminar, were dedicated to the quest of the historical Jesus. These scholars questioned the historical authenticity of most of the traditions about Jesus recorded in the New Testament gospels. From their method of analysis they conclude that there is no internal evidence in scripture to support Papias's report that Mark's gospel is anything close to a transcription of what he had heard from Jesus's disciple Peter. A host of other scholars employ methods of historical analysis that lead them to be much more trusting of the traditions attributed to Jesus in the gospels.[12]

A simple comparison of Mark's gospel with the gospels according to Matthew and Luke shows that the latter two evangelists adapted what they heard from Mark to the needs of their listeners. Papias says that Peter used this technique of adaptation from the very start. If Mark's gospel is based on what he heard from Peter, he followed the same practice as he composed his artful and nuanced narrative. In other words, we do not have in Mark or any other gospel a blow-by-blow account of Jesus's life or verbatim transcriptions of what Jesus said. The traditions were adapted to the needs of the audience.

## HEARING MARK IN CONTEXT

Mark told his story in the hope of guiding his listeners to a new understanding of where history in this world was taking them and how they could live their lives in partnership with God's goal of healing the world God created. In addition to knowing the background story of Israel's cosmic narrative to which Mark alludes, it is important to grasp how his story unfolds. Mark's narrative follows a unique course of plot development, moving from the opening of Jesus's ministry to his death and his resurrection. A plot that ends in rising from the dead would be strange enough in itself. But as we shall see, according to what was most likely Mark's original ending at 16:8, the witnesses to his resurrection run away in fear and don't tell anybody about it. This leaves the attentive listener to wonder how Jesus's vision of his followers preaching the gospel to the nations (13:10) will be realized. Not only that, but after an initial period of roaring success in healing people and amazing crowds, Jesus prophesies that at the showdown with the authorities in Jerusalem he will die without even a fight. That is a plot first to perceive and then to ponder.

Mark's compositional techniques may also be unfamiliar to listeners and readers, making it difficult to follow his story in anything more than a superficial way. His narrative exhibits many of the characteristics of oral storytelling identified by scholars over the past half-century. These include the aggregation of small episodes arranged by theme: stories of his beginnings, his Galilean ministry, his journey to Jerusalem, and his passion, death, and resurrection.

The deliberate arrangement of the episodes within a theme creates discernible patterns. For example, five stories of Jesus's healings are followed by five stories about controversy. One episode is sometimes enclosed within another to create a relationship that Mark does not specify but leaves to the listener to discern. We can see simultaneity when the anointing for burial is enclosed between the search by the authorities for a stealthy way to arrest Jesus and the arrival of Judas to solve their problem. But the pairing of the two stories also generates irony, the same with the pairing of Peter's denials and Jesus's trial.[13]

Verbal threads are another technique. These evoke associations and are strategically placed ("my beloved Son" at the transfiguration echoes the same phrase at the baptism). This is just a sample of Mark's compositional techniques that generate meaning and also aid the memory of the oral storyteller.

Compared to the gospels of Matthew and Luke, Mark's spare, elegant narrative has been regarded in some scholarly circles as a bare-bones, "just the facts" version of the life of Jesus. K. L. Schmidt, an early twentieth-century German scholar, compared the structure of the Gospel of Mark to "a heap of pearls, some of which are connected to each other."[14] But there is more to Mark's narrative. In his study contrasting Homer's story of Odysseus's scar in the *Odyssey* and the story of the sacrifice of Isaac in Genesis 22, Erich Auerbach notes that in contrast to Homer's story, the story in Genesis is "fraught with background."[15] That same term can be applied to Mark's narrative. Many background details are missing—for example, with what specifically did Satan tempt Jesus?—and the impulse is for listeners to fill the gaps. Mark's story is saturated with allusions to Israel's traditions that the audience has to "get" if they are to perceive the fuller significance of the little that Mark does tell.

Paying close attention to Mark's story and the larger narrative told or implied in the traditions of Israel prepares us to see new meaning in the rest of the works gathered in the New Testament. Matthew and Luke take up Mark's narrative line and most of his individual episodes and teachings; they augment the story he tells with many additional teachings as well as some other narratives, which they frequently modify. They are also laden with allusions to Israel's traditions. Mark's story raises questions about the larger narrative world behind the Gospel according to John. This gospel follows a different plot line and recasts earlier traditions about Jesus into new genres and rhetorical forms of expression. It also alludes to Israel's tradition and is widely understood by scholars to have been performed for an audience of Israelites living in a culture heavily influenced by Greek philosophy and rhetoric.[16] A new understanding of the letters of St. Paul also emerges when we listen to them from the perspective of what we've learned about Mark's story of Jesus. For most of Christian history people have heard and quoted Paul's letters selectively in order to fit a story of the cosmos different from the Israelite story that Paul assumed.[17] The story that unfolds in the Book of Revelation also sounds different when heard in relation to the story of Jesus that Mark tells and the Israelite story of the cosmos that Jesus's story brings to completion.[18]

Mark speaks powerfully to concerns for the overwhelming problems we face today—violence, poverty, injustice, and environmental degradation. His primarily Israelite audience was composed of people committed to solving their problems rather than escaping from them into some other spiritual world. The Israelites were astute observers of nature and human nature. In attending to the material world in which they lived an embodied life, they were grounded in a deeply spiritual, religious perspective. They asked the existential questions we ask today: What are the origins of this world? What is my place in it? Is the force that brought us into being still active? How can we be a force for good?

The Gospel of Mark and the Hebrew scriptures shed considerable light on our current spiritual, moral, and environmental crises. The curse that God pronounced on the first humans and their descendants in Genesis describes the origins of our disrupted human relationships and place in nature. It relates the subordination of women to the power of men and to the introduction of violence, death, and discord on Earth. Mark's story tells how the life and death of Jesus lifted that curse, healing a broken world and inviting us to participate in that healing in our own times. In that story is meaning and metaphor that usher in a new way to achieve unity with the God who created our material world and filled it with a dazzling variety of creatures. God finally created us humans and invites us to share the joy of all creation, our original and enduring home.

Contemporary scholars and theologians who recover the original meaning of Mark and other scriptures enable us to take more seriously the spiritual aspects of the challenges that lie before us. As their work becomes more accessible, people of all faiths can discover the riches of modern scripture scholarship and the wisdom of cosmological theology. As we devote ourselves to generating fields of compassion, we can all participate in the healing of human brokenness and the transformation of our world.

# 3

# "He Was Among the Wild Beasts"

The best-known story of the temptation of Jesus in the wilderness is the version told by Matthew and Luke, full of dialogue and vivid details. In Matthew's version, Jesus, famished after a forty-day fast, engages in a dialogue with the devil, who entices him to abuse his power as the Son of God. First, he taunts the hungry Jesus to turn stones into bread. Then he dares him to defy death itself: leap off the highest tower on the great Temple of Jerusalem, trusting in God's angels to save him. The devil's final deal is a business transaction: he entices Jesus to sell his soul in exchange for wealth and worldwide power.

Luke reverses the order of the last two anecdotes. But each evangelist saves for the last the temptation that resonates with an important theme of their respective Gospels.[1] In both narratives, Jesus wards off temptation with citations of Hebrew scripture, and the defeated devil leaves the scene. By resisting the devil's taunts to use his power to serve his own interests, Jesus reinforces his calling to inaugurate the rule of God.

Most scholars believe that Matthew and Luke used Mark's account as a basis for their narratives. But Matthew and Luke also added material—the temptation story, among others—drawn from a common source, which scholars have dubbed "Q," consisting mostly of oral narratives of the teachings of Jesus.

To label this common source "Q" may seem like drab scholarly packaging for a rich trove of stories about the teachings of Jesus. But Q, drawn from the German word *Quelle*, meaning a source, is a metaphorical spring that serves as the headwaters of a mighty river. Reflecting on the account of Jesus's temptation in Q can become the source of insights about his actual ministry in the brief period of his public life. These temptations are the source of a great irony of his mission. Although Jesus rejects the devil's temptation to throw himself off the pinnacle of the Temple, his radical message and rising popularity in a repressive political environment ensure that Jesus courts death throughout his public life, trusting that God will care for him and raise him from the dead. At the end of Matthew's Gospel from a mountaintop he declares, "All power in heaven and on earth has been given to me. Go, therefore, and make disciples of all nations." The irony here is that he has gained all the kingdoms of the world offered him by the devil—but after a life of serving God, not the devil. And while he will not turn stones into bread to feed himself, he will twice turn five loaves and two fish into bread to feed multitudes.

Mark's terse account at first pales in comparison to Matthew and Luke. It provides no details about the actual temptation. Mark simply says, "He remained in the desert for forty days, tempted by Satan" (Mark 1:13). The rest he leaves to our imagination. Mark does not mention that Jesus was fasting. Nor does he explicitly indicate that Jesus successfully resisted temptation. This is characteristic of Mark's storytelling style. The absence of information invites us to stop, think, and raise questions.[2] The next two scenes, expressed in one brief sentence, give us cause for deep reflection, as we try to puzzle out both the sense and symbolism:

> He was among the wild beasts, and the angels ministered to him.

Who or what are these wild beasts? What do they represent? What was the nature of the interaction? What role do these manifestations of nature play in Jesus's temptation or its aftermath? Why pair an image of the wildness of nature with the ethereal presence of angels? Are they there to comfort him in the loneliness of the desert night? Or is their comfort a reward for resisting temptation unassisted? Mark's narrative leaves those answers entirely up to the listener. It is a fruitful meditation to consider what this all means.

Biblical commentaries on this brief passage vary widely. In a footnote, the *New American Bible, Revised Edition* offers two completely different interpretations: "The presence of the wild beasts may indicate the horror and danger of the desert regarded as the abode of demons or may reflect the paradise motif of harmony among the creatures; cf. Is 11:6–9." *The HarperCollins Study Bible* calls attention to the peaceable kingdom of animals living in harmony depicted in Isaiah 11:6–9 and the scene of Adam at peace with the wild beasts in Genesis 2:19–20: "[I]t may suggest the restoration of a paradisiacal condition that existed before the Fall."[3] And *The Oxford Annotated Bible* makes no comment about the wild beasts at all or indeed of any connection with the story of the temptation in Genesis.[4]

Israelite listeners to Mark's spoken story in the century after Jesus's death were familiar with Genesis; they would recognize that Jesus's temptation by Satan was a replay of Adam and Eve's encounter with the serpent.[5]

A vision from the prophet Isaiah, when remembered in the context of the story of Adam and Eve, suggests a cosmological significance to Mark's story of Jesus encountering wild beasts. Isaiah envisions a time in the future when violence between humans and beasts or among beasts themselves will no longer be a part of life in this world. This will happen, according to Isaiah, when God places on the throne of Israel a king who will be filled with the spirit of God and, like Adam and Eve before the Fall, will rule the Earth according to God's will. The words describing this vision were first spoken about eight hundred years before Jesus. From the time Isaiah and other Hebrew prophets were put in writing and gathered into a second collection of "scriptures" alongside the Torah, people have been able to hear or read again their oracles. Today Isaiah 11:6–9 is viewed by some biblical scholars as a vision symbolizing a renewed harmony with all creation:

> Then the wolf shall be a guest of the lamb,
> and the leopard shall lie down with the young goat;
> The calf and the young lion shall browse together,
> with a little child to guide them.
> The cow and the bear shall graze,
> together their young shall lie down;
> the lion shall eat hay like the ox.
> The baby shall play by the viper's den,
> and the child lay his hand on the adder's lair.
> They shall not harm or destroy on all my holy mountain;
> for the earth shall be filled with knowledge of the LORD,
> as water covers the sea.

Together, Genesis and Isaiah tell the story of an original harmony between humans and beasts (Genesis 2:18–20), followed by a disruption of that harmony as a consequence of the first humans' rejection of God's will (Genesis 3:15). This, in turn, is followed by a vision of a restored harmony between humans and animals (Isaiah 11:6–9; also 65:25[6]). In this context, Mark's brief report, "He was among the wild beasts, and the angels ministered to him," signals right from the moment Jesus withstands Satan's temptation that Paradise is being restored.[7]

## WRESTLING WITH TEMPTATION

Throughout history humans have had to wrestle not only with physical threats in the natural environment but also with their internal passions and demons. The wild beasts signify both a physical threat and a powerful metaphor for a whole range of human impulses and temptations with which Jesus must have struggled.

As a fully formed human being, Jesus must come to terms with the savage side of his nature,[8] which would instinctively kick into gear when faced with danger. Facing his own demons, he is free, as were Adam and Eve, to obey or disobey God's will. The temptation to disobey would remain with him through his entire earthly life. This is an identifiable narrative tension that threads its way through Mark's story.

Immediately after his initial temptation, Jesus comes into Galilee announcing the inauguration of the Kingdom of God upon the Earth. Clearly, in Mark's narrative, a new Adam has arrived in the world; his unbroken harmony with the Creator gives him the power to lift the primeval curse brought about by Adam and Eve's disobedience. His concrete actions manifest the ending of the curse.

Jesus's initial success against Satan in the desert does not end his struggle to remain faithful to God. Halfway through Mark's narrative, after an intense period of healing and exorcizing demons and welcoming sinners and engaging in disputes with the religious leaders, Jesus clearly articulates the hard realities that await him:

> He began to teach them that the Son of Man must suffer greatly and be rejected by the elders, the chief priests, and the scribes, and be killed, and rise after three days. (8:31)

In this scene, Mark's character of Peter tempts Jesus, rebuking him for choosing a path of certain death. Jesus, in turn, rebukes Peter:

> Get behind me, Satan. You are thinking not as God does, but as human beings do. (8:33)

The struggle against temptation strongly suggests itself again toward the end of Mark's story. In the poignant narrative of Jesus's prayer in Gethsemane, the crucial moment of decision to either surrender to his enemies or escape, he prays to his Father to release him from the need to suffer and die. In Mark 14:38 he exhorts Peter, James, and John:

> Watch and pray that you may not undergo the test. The spirit is willing but the flesh is weak.

The Greek word the *New American Bible, Revised Edition* translates as "undergo the test" is the cognate for the verb Mark uses in 1:13 for "tempted." These words are a warning to the disciples, but they seem to emerge from Jesus's own struggle between his spirit and his flesh. In setting up the story of the prayer in 14:34–35, Mark gives one of his rare explicit descriptions of how Jesus is feeling about it. The *New American Bible, Revised Edition* translates Mark's Greek:

> [He] began to be troubled and distressed. Then he said to them, "My soul is sorrowful even to death. Remain here and keep watch." He advanced a little and fell to the ground and prayed.

Some other translations employ more emotionally intense language. *The New Jerusalem Bible* says:

> And he began to feel terror and anguish. And he said to them, "My soul is sorrowful to the point of death. Wait here, and stay awake." And going on a little further he threw himself on the ground and prayed.

*The Revised English Bible* translation is similarly forceful:

> Horror and anguish overwhelmed him, and he said to them, "My heart is ready to break with grief; stop here, and stay awake." Then he went on a little farther, threw himself on the ground, and prayed.

This description of Jesus's feelings obliges an oral narrator to enter into Jesus's experience and speak his words to God with great anguish. Jesus bows to the will of God. This comes only after an outpouring of intense pleading, followed by a long pause during which Jesus regains his breath and settles down in his spirit to the point of accepting God's will. This is the ultimate temptation a human being can face: to choose to save his life rather than obey the will of the one who gave him life. God is asking Jesus to face down death, giving it no power to determine his actions, trusting in God to raise him from the dead to a life like that originally enjoyed by Adam and Eve. By obeying the will of God, Jesus, ironically, essentially agrees to do what the devil tempted him to do in the story Matthew and Luke got from Q. To go willingly to the cross in hope of resurrection is as daring as a leap off the pinnacle of the Temple in hope that angels will bear him up. But Jesus makes his death-defying plunge into the abyss in obedience to God, not to the devil.

Temptations are objects or actions to which we are powerfully drawn. Under the right circumstances, temptations can be good for us; under different circumstances they can do us real harm. We may experience temptations as hard-to-tame impulses. In the realm of psychology, such impulses can take the form of delusions, hallucinations, and intrusive, dangerous, or uncontrollable thoughts. Like raging beasts, these forces threaten to overwhelm and ultimately destroy our sanity.

Anyone who has a stated goal or mission in life faces temptations to be or do things that will get in the way. What were Jesus's temptations? Matthew, Luke, and Mark all say that at his baptism Jesus heard a voice from heaven declaring that he was God's beloved Son. What would it have involved for Jesus to remain faithful to that identity and what would have gotten in Jesus's way?

## SEXUAL TEMPTATION

The temptations in the stories told by Matthew and Luke suggest Jesus was grappling with different ways his power could be misused. Mark, in contrast, makes no suggestions about the nature of the wild beasts that might have been tempting Jesus. We are free to imagine what they might represent for a man in Jesus's position. We might wonder, for example, whether he struggled with the role human intimacy would play in his life.[9]

To the great fourth-century church leader Ambrose, who was Bishop of Milan and mentor to Augustine of Hippo, it was incomprehensible that Jesus had any sexual feelings at all. In Ambrose's culture and among the philosophers and theologians of his day, the sexual impulse was a stain on human nature, the result of the fall of Adam and Eve. Ambrose's antipathy about sexual feelings may have been in response to the predatory and violent sexual norms of Roman society. As an Israelite, Mark would not have shared that negative view. While there are stories of sexual violence in Israel's scriptures, they stand out in contrast to the ideal of sexual fidelity commanded in the Torah. From the stories of Genesis to the Song of Songs, the exquisite passions of intimacy are evoked as an image for both an ideal for human intimacy and for the relationship between the chosen people of Israel and their God.

Jesus was a fully human person; he experienced the same passions, yearnings, impulses, and conflicts as any other mortal. He surely had to struggle with the role sex and the appeal of deep human connections would play in his life in relation to his vision of how he would carry out his vocation as obedient Son of God, whose destiny was to lift the curse of Genesis and restore us to harmony with all creation.

This does not mean that for Jesus intimate expressions of love for another person were categorically bad, as so many theologians and religious leaders have thought through the ages. It only means that there are different ways people choose to live out their roles as sexual beings. Some sexual experiences are unquestionably harmful and exploitative; others are a source of great blessing. But even a joyful and responsible experience of intimacy and sexual fulfillment might not be an option if it gets in the way of a goal or a mission that a person wants even more.

In the thirteenth century, Thomas Aquinas spoke of the way our primary desires can get in the way of secondary or higher desires.[10] Mark's narrative of how Jesus recruits his disciples makes it clear that for him the responsibilities of being a husband and father would conflict with what he needed to do to bring the rule of God into the lives of others. Those men he called to be among his twelve closest disciples were asked to leave their families in order to follow him. But this does not necessarily mean that they abandoned their loved ones. One of Jesus's first healings is of Peter's mother-in-law. That means he must have had a wife, with whom he had a deep relationship, possibly with children. Gerhard

Lohfink points out that Jesus and his circle of twelve who freed themselves to travel with him depended on the support of settled families sympathetic to his mission for food and shelter.[11]

## VIOLENCE AND OTHER PSYCHIC BEASTS

Mark's story of Jesus in the desert introduces the notion that Jesus also struggled with other dark forces that reside within every human psyche—fear, jealousy, greed, anger, hatred, violence. The temptation to be afraid in the face of opposition or to employ violence to overcome opposition would also have been a significant issue for Jesus.

Violence has been the default choice for humans since the Fall when, in pursuing our goals, we meet with violent opposition. Israelites of Jesus's day who were looking forward to replacing the rule of the Romans with the rule of God looked for a warrior messiah like the great King David to lead them.[12] When did Jesus resolve that he would not lead Israel in a violent revolt as the zealots did a few decades after Jesus's crucifixion? We can only imagine that it must have been during his forty days in the desert when he was figuring out how to replace the rule of Satan in the world with the Kingdom of God. Peter would have expected Jesus to enter Jerusalem as a warrior who would vanquish his enemies with overwhelming violence as King David did. Jesus says to Peter, "Get behind me, Satan!" Victory through violence is the way of human beings, not the way of God.

The power of wild animals as a metaphor for Jesus's inner struggle with temptation does not exclude their meaning on a literal level. Mark and his listeners lived in an environment where attacks from animals were a constant danger. Indeed, if you were not attacked while traveling in open country, the Romans brought the spectacle of animal-on-human violence to your local theater—the Coliseum—where it was used either to entertain bloodthirsty crowds or to end a person's life altogether, for breaking Rome's law or following the teachings of a radical preacher.

In our own time, humans maintain a love/hate relationship with the natural world. We love our pets, we marvel at the night sky, and we thrill to scientific discoveries, yet we also have much to fear from nature: wild animals, wildfires, destructive storms, epidemics and killer viruses, or rat-infested houses of the poor. As much as humans are at ease with nature, we are also willfully destructive of it: We hunt animals for sport and subject them to painful medical experiments; we condemn cattle, pigs, or chickens on factory farms to miserable lives in filthy, crowded conditions.[13] Our thirst for fossil fuel drives us to molest the earth with fracking and oil wells, with the consequential earthquakes and ruin of pristine aquifers. We modify the genomes of plants and animals without fully knowing the genetic outcome. We drive cars that pump poisons into the atmosphere. We pollute our waters with sewage and industrial waste. Are we any less barbarian in our exploitation of the natural world than the ancient Romans were with the beasts they let loose in the Coliseum? The phrase "crimes against humanity" is a term used in international law. The concept "crimes against nature" has yet to become a legal term. Invoked against modern civilization developed in the United States and Western Europe, it would indict us not only for many of our practices but for our basic attitude toward the natural world.[14]

## JESUS AND NATURE

The palpable transformation of God's suffering creation is embedded in the mission of Jesus. And we can clearly see in Mark's story that his work included the world of nature as well as human bodies, spirits, and society. Mark's opening image of Jesus as a human enjoying the restored harmony with the wild beasts Adam knew in Paradise signals the healing of the human relationship with all of nature. Having tamed the inner beasts that might draw him away from oneness with God, he is able to extend healing to the broken relationship between humans and nature.[15]

One of the clearest manifestations of this realm of healing occurs when Jesus and his disciples find themselves in a boat tossed about on stormy seas. Jesus, completely at ease in nature, appears to sleep through the storm. But when awakened by their distress, he calms both the waters and his companions' fears.[16]

The sea is for the Israelites a fundamental symbol of chaos. The story of the creation of the universe in Genesis begins with a picture of watery chaos that God then proceeds to bring into order:

> and the earth was without form or shape, with darkness over the abyss and a mighty wind sweeping over the waters. (Genesis 1:2)

Commentators frequently see in the disciples' experiences on the turbulent sea a metaphor for the spiritual turbulence into which their lives have been thrown since they have decided to follow Jesus. Rarely do commentators wonder whether Jesus's concern to quell the destructive forces of nature calls us to do what we can to prevent, minimize, or remedy the effects of nature's destructive potential. What human practices cause nature to be destructive or exacerbate that potential? We may have a difficult time believing that Jesus actually calmed the wind and the sea. But why limit our interpretation to the calming of our emotional and spiritual turmoil? Why not also ask about our role in both generating and doing what we can to minimize and even prevent natural destruction?

We know that from the time of Augustine, Christian tradition in the West has emphasized the inner spiritual relationship between individuals and God. Our consequent detrimental relationship with the Earth and the vision of a transformation that would include the Earth dropped out of awareness. The goal after death is to depart the material world and spend eternity with God. This limitation is exacerbated in the emphasis in Protestant theology on salvation by faith alone, rather than good works one might do in the here and now. Protestant leaders like Martin Luther and John Wesley were clear in saying that works were the fruit of faith,[17] but that emphasis has largely been ignored by Evangelical Protestant Christians today: There is a widespread aversion to the idea that Jesus had any interest in contemporary economic, political, social, and environmental issues other than issues connected with sex. Christians have largely become deaf to the meaning of Jesus's mission for the healing of the material creation itself.

The urgency of our current environmental crisis—climate change, extreme weather, pollution, and mass extinctions of species—warns us not to ignore the importance of the story of Jesus for our responsibility to our natural environment. The relentless focus on the afterlife has blinded us to our responsibilities to preserve and protect the environment and the spiritual richness of the natural world. But there is another rationale for our carelessness about creation, a quirk in nineteenth-century American theology that has taken hold in popular culture and complicates the task of responsible stewardship of the Earth.

## DISPENSATIONAL THEOLOGY

In the nineteenth century, a dystopian worldview far different from the peaceable kingdom of Isaiah's prophecy took hold of the Christian imagination: a doomsday scenario of an evil world full of misguided people who are eventually destroyed by an angry God. The few righteous souls who escape destruction are "raptured," taken up into heaven for a life of eternal bliss. It originated with John Nelson Darby, an Irish Anglican who was disenchanted with the condition of his church.[18]

Darby's theology is known as "dispensationalism," the view that the Bible divides the history of the world into a succession of "dispensations" or eras, each with its own rules. Darby cobbled together a scenario of the future of the world by picking and choosing passages from the Hebrew prophets, the New Testament, and especially the Book of Revelation. Darby's doomsday scenario excised scriptures from their cultural and literary contexts. Incorporating them into his new scheme radically transformed their meaning.

Dispensational theology suggests that all the problems we face on Earth can only be resolved in a fiery destruction and there is nothing anyone can do to avoid it. This highly selective and defeatist interpretation of the Old and New Testaments determines the worldview of many conservative Christians—as well as non-religious people who have reason to be pessimistic about the future.[19] The flip side of the dispensationalist worldview is seen in the false promises of the Prosperity Gospel, which equates poverty with sinfulness and wealth as evidence of divine favor.

In the dispensationalist view, any attempts to address the tribulations we are experiencing in our world are misguided. World peace will only be achieved when we humans are on the threshold of completely destroying ourselves, at which point Jesus will literally return to the Earth to establish a thousand-year reign known as "the millennium." In the meantime the only hope and task for Christians is to believe in Jesus Christ as Savior, study the prophecies, and be among those ready to be secretly "raptured" by Jesus to safety in a place he has prepared. After his thousand-year reign on Earth, all the atoms that make up material reality will be taken apart. Then Christ will reassemble them to construct a New Heaven and New Earth that, unlike our own world, will not be subject to corruption.

Dispensationalist theology divides humanity into the few righteous and the masses who are damned, which includes not only non-Christians, but Christians who are not studying the prophecies. They see war as the will of God for bringing us to the verge of self-destruction. They see no point in resisting that scenario, since it is scripturally ordained. In fact, many dispensationalists encourage war. They see no point in trying to reverse our damage to the natural environment, because it is doomed.

Darby's dark vision gained steam with the publication of the *Scofield Chain Reference Bible*. Published when the world was on the brink of World War I and revised in 1917, a year before the war ended, it reflected the anxieties and helplessness of the modern era. The *Scofield Chain Reference Bible* functioned as the authoritative text of a highly effective social network on multiple platforms. By pairing the complete King James version of the Bible with a cross-referencing system of notes and apocalyptic commentary that selectively interpreted biblical passages from an apocalyptic perspective, the *Scofield Chain Reference Bible* fueled a social network of its day, cultivating a robust cohort of fundamentalist Christians steeped in dispensationalist theology and fundamentalist interpretations of scripture.

Beginning in the 1950s, American televangelist and novelist Hal Lindsey followed in Darby's footsteps, amassing a large following of believers who were captivated and terrified by *The Late, Great Planet Earth*.[20] Later, televangelist Tim LaHaye and co-author Jerry Jenkins continued the trend with their *Left Behind* series of novels. The appeal of this worldview endures today among self-identified Christians who prefer predetermined judgments and a punitive gospel of fire and brimstone to the compassionate healing actions of Jesus. This "all is lost" thinking also finds a home in the apocalyptic plots of movies and television series, which are often completely detached from the Christian tradition.

The view that originated with Darby is today aggressively promoted and eagerly accepted by millions of believers as faithful to the Bible and to Jesus. It is the basis for the right-wing political views of many Christians in America, who resist efforts to save the natural environment, press for a policy of unrestricted support for the state of Israel's violence against Palestinians, see Muslims as the new enemy on whom to make continual war, and balk at any strategy aimed at reducing violence other than more violence.

The dispensationalist interpretation of the Bible is completely at odds with our reading of the Gospel of Mark. It does not conform to the narrative of God's activity and vision for the restoration of creation in Israel's scriptures in which Mark's story is nested. Absent from it is Jesus's announcement at the beginning of his ministry that with deeds and words the Kingdom of God is beginning to replace the rule of Satan and his demons upon the Earth. Absent is Jesus's embrace of outsiders like tax collectors and the Gentiles. Absent is his assertion that the Temple is a place for the Gentiles to pray with Israel to their God. Absent is Jesus's ministry of healing sick and maimed human bodies and feeding the hungry with an abundant feast.

Dispensationalism fails to acknowledge our collective responsibility to tend and care for God's good Earth and to join with Jesus to make the Kingdom of God a concrete reality through acts of compassion and healing.

In Mark 13 Jesus shares a vision of the future with his disciples. He first prophesies the destruction of Jerusalem's magnificent Temple, then unfolds a vision of upheaval and suffering: conflict among nations, natural disasters, persecution of the disciples, and people making false claims to be the messiah. The suffering will engulf the farthest reaches of the cosmos, but in the end Jesus the Son of Man will triumph. This vision may at first seem to support the dispensationalist conviction that cosmic cataclysm is a necessary piece of the divine plan. The images of cosmic cataclysm are drawn from several of the Hebrew prophets, a fact that further suggests that God wills cosmic destruction.

What Jesus is foreseeing here, however, is not the doing of God's will but the consequences of ongoing human disobedience under the influence of Satan. The resistance to the further spreading by Jesus's disciples of the good news of the reign of God and defeat of Satan (Mark 13:9–11) will be a continuation of the resistance experienced by Jesus that culminated in his crucifixion. For the next three centuries after Jesus until Constantine embraced Christianity as the favored religion of the Roman Empire, faithful followers of Jesus did in fact suffer persecution for living a life that was being delivered from the curse of Eden. Roman life was structured around all the characteristics of that curse. Paul's letters, the Book of Revelation, and other early Christian works testify to the price followers of Jesus paid for living in a way that challenged the prevailing order.

Even when Jesus does foresee war and environmental ruin, he envisions his followers remaining faithful to his mission. There is not one suggestion in the entire New Testament that Christians are to aggravate those conditions or rejoice in them. When we look at Jesus's mission during his earthly life, we see him resisting them, bringing healing in their place, and teaching his disciples to join in this curse-ending work.

All the world's turmoil that Jesus foresees will end with the coming of the Son of Man. He prophesies this event three times in the course of Mark's story: 8:38, 13:26, and 14:62. While the term "Son of Man" is found in several Israelite compositions, including the Hebrew prophet Ezekiel and a very popular corpus of non-scriptural works bearing the name Enoch, the picture of the Son of Man coming with the clouds of heaven is clearly drawn from Daniel 7:13. In this vision of Daniel, the empires that have ruled Israel since the time of the Babylonian Exile are represented as beasts, an appropriate symbol given their ravenous and murderous treatment of all those they had conquered. In Daniel 7 these beastly powers will be defeated, and their rule will be replaced by a Child of Humanity, the Son of Man.

In *Jesus and the Victory of God*, N. T. Wright argues that the vision articulated in Daniel 7 was widely popular among Mark's Israelite contemporaries.[21] Many of them found in it the inspiration to rise up in violent revolt against Rome. Those who listened to Mark's story and heard Jesus speak of the Son of Man coming on the clouds would recognize the allusion to the "one like a son of man" in Daniel. Hearing it in Mark's context, they would find in it a very different significance. For Jesus this "one like a son of man" would bring to fulfillment Jesus's mission of reconciliation that included the Gentiles as well as all Israelites.[22] In Daniel 7 God gives this figure dominion over all the nations. His rule on God's behalf replaces the rule of all the beastly empires. Gerhard Lohfink[23] points out that the rule of "one like a son of man" is a humane rule, exercised without any form of violence. This is the Son of Man who "did not come to be served but to serve and to give his life as a ransom for many" (Mark 10:45).[24]

Mark's story ends with the apparent defeat of Jesus and his mission, not its success. But in the vision he shares with his disciples on the Mount of Olives just before his crucifixion (Mark 13), he envisions his disciples preaching the gospel until it reaches all the nations of the Earth. Jesus's words contain no hint of dispensationalist thinking. They contain no reference to the world ending in a final conflagration. He does not envision his disciples being taken away in a rapture, leaving suffering humanity behind. If they are faithful to the mission of

Jesus, they would be far more likely to be the targets of Satan's assaults. Their joy is to be found in the healing and peace they bring to the world.

## GRACE AND NATURE: A NEW NATURAL THEOLOGY

Cosmological theology is deeply committed to the health of planet Earth as the arena of God's saving work and therefore of Christian responsibility and possibility. Cosmological theologians seek to join the Christian narrative of Jesus's part in saving the world to the story of the Earth's evolution emerging from modern science and to the implications of scientific discoveries.

The Jesus we meet in the gospels of Mark, Matthew, and Luke clearly wants people to trust him—but for what? As Mark tells the story, until his trial before the high priest and his death, the only ones who do name him the Son of God, besides Mark in his opening line and God at Jesus's baptism and transfiguration, are the demons he commands to keep silence. Jesus calls people to do what he is doing. They surely need to trust that his message is the way of God for the life of the world. He calls them to put his teachings into action.

Pope Francis has made care and protection of our fragile planet a moral issue, not only for Catholics, but for all people. Christians could help lead the way, for the sake of all humanity.

The vision of salvation in cosmological theology is one that includes all creation. Jesus's harmonious relationship with the animal world, his command over the metaphorical seas of chaos, his healing of broken bodies, and his feeding of hungry people exhibit the way human life and the health of the natural environment are bound together, as they clearly are in the myth of our beginnings as told in Genesis. Cosmological theology seeks to make this vision of salvation the default theology of the Church. But that calls for significant change.

In *Christ in Evolution*,[25] Ilia Delio explores how theologians Joseph Sittler, Jurgen Moltmann, and Zachary Hayes are among those looking at new ways to expand our understanding of redemption, from personal morality to a dynamic engagement with the natural world:

- The Western church's distinction between nature and grace led to a contempt for nature (50). . . . We must now take our stand, he [Sittler] said, not with Augustine but with Irenaeus, for whom all goodness, whether in this world or at the final consummation, manifests the grace of God. . . .

- Moltmann claims that, in light of Sittler's address, we must confront the "threat to nature" with a "Christology of nature" in which the power of redemption does not stop short at the hearts of men and women and their morality but extends to all of nature. (51)
- Franciscan theologian Zachary Hayes notes that the understanding of the doctrine of the incarnation from Alexander of Hales to Scotus, including Bonaventure, "did not limit the discussion of the meaning of Christ to the reality of the cross" but expanded it to the widest possible horizon. What these theologians did, he said, is to "perceive the possible relations between the story of Jesus and the larger picture of the world." (56)

Just how limited is our world to currently known possibilities? Time after time, human beings have not been satisfied to accept things the way they are and to believe that we can do nothing to change them. In fact, the history of the evolution of our world through the agency of the human imagination is the history of effecting significant change. The question remains whether we have the desire and the commitment to devote our imaginations, our energies, and our resources to generating fields of compassion strong enough to overcome the wild beasts of fear, greed, pessimism, and doomsday thinking that are ravaging our souls, our psyches, our planet, and all creation.

# 4

# Jesus and Women

For a long time, scripture scholars have regarded Mark's gospel simply as a stripped-down, just-the-facts account of the public life of Jesus. It is the earliest of the gospels, and the later accounts of Matthew and Luke both draw from Mark, adding stories of Jesus's birth, resurrection, and teachings, including the most well-known and loved parables. Their language is more literary in style, especially Luke's. Matthew often explicitly quotes Israel's scriptures as prophecies fulfilled in Jesus, although all three gospels are saturated with allusions to the scriptures.

Mark's narrative is terse in comparison. His style is minimalist, cinematic, and dramatic, giving the story a powerful and memorable impact when performed before a live audience. The century after Jesus's death was a time of religious, cultural, and political turmoil, of exploitation and brutality. This story was likely told in public places, in synagogues and private homes,[1] to an audience primarily of Israelites, who were trying to make sense of their lives in a troubled time. Would this Jesus give them new ways to draw closer to God so that the world would be transformed?

Biblical scholars are only now beginning to appreciate that Mark's story is deceptive in its simplicity. He takes a documentary approach, bearing witness to the public life of Jesus in his journeys across the parched landscape of first-century Palestine and his encounters with its diverse inhabitants in homes, in public squares, and on the dusty roads of Galilee. The story unfolds scene by minimalist scene,[2] artfully arranging and weaving together Jesus's inspirational teaching with episodes of healing and controversy, reconciliation and conflict, incomprehension, explanation, and correction.[3] It climaxes with a brutal account of betrayal, torture, and execution. And though ancient scribes in subsequent centuries tacked on a more emotionally satisfying ending, the original conclusion of Mark's gospel is a clear-eyed acknowledgment of human brokenness, lit with flashes of courage, compassion, and hope.

Mark's story provides a picture of Jesus's relationship with his closest followers as they accompany him on his redemptive journey. Those intimate friends fall into two broad categories: the twelve men he chooses as his apostles and the uncounted number of faithful women, most of them nameless, who accompany Jesus, embrace his teaching, support him, care for him, love him. And after Jesus releases his last, agonized breath, Mark's cinematic narrative zooms in to a close-up of those who stayed with him to the end, long after the privileged males who were his chosen disciples had abandoned him:

> There were also women looking on from a distance. Among them were Mary Magdalene, Mary the mother of the younger James and of Joses, and Salome. These women had followed him when he was in Galilee and ministered to him. There were also many other women who had come up with him to Jerusalem. (Mark 15:40–41)

This brief but compelling image demonstrates that women comprised a significant number of Jesus's first followers. In Mark's telling of the story, they sound like little more than an afterthought. But Mark's gospel makes clear that Jesus had deep, meaningful, and authentic encounters with women, interactions that not only challenged the patriarchal culture of the times, but also signified one of Jesus's most urgent redemptive missions: to lift the curse in Genesis that rendered women inferior to men, a scriptural pronouncement that has long been used to authorize the subjugation of women in the Church and in the world.

## A BREAK FROM PATRIARCHY

Richard A. Horsley, Elizabeth Struthers Malbon, and other scripture scholars are finding complex and revealing patterns in Mark's narrative that cast new light on Jesus's encounters with women and the leadership roles he welcomes them to fill.

In his 2001 book *Hearing the Whole Story: The Politics of Plot in Mark's Gospel*,[4] Horsley looks at Jesus's encounters with women in the context of the patriarchal structure of the culture in which he lived. Patriarchy was not peculiar to Israel's culture, nor was it essential to Israel's religion.[5] Male dominance, nevertheless, was a fact of life in Jesus's day. Women were identified mainly in terms of their connection with men. They generally did not associate in public with men who were not members of the family. And it was taboo for women to make overtures to men outside the family circle. On all three counts, Horsley writes, over the course of Mark's narrative a pattern emerges of women increasingly acting and being named in ways that challenge the patriarchal norms of Israelite society of Jesus's time.

There are eight episodes in the Gospel of Mark in which Jesus's life intersects significantly with women. They span the gospel from one end to the other:

- Jesus cures Peter's mother-in-law (1:29–31).
- He raises the daughter of Jairus from the dead (5:21–24, 35–42).
- He cures a woman suffering from a hemorrhage (5:25–34).
- He encounters a Gentile woman and cures her daughter, traversing cultural and ethnic boundaries (7:24–30).
- He debates the Pharisees on the permanence of marriage, equality of the sexes, and legality of divorce (10:1–12).
- He praises a poor widow who gives her meager savings to the Temple treasury (12:41–44).
- He accepts the intimate ministrations of an unnamed woman in Bethany. She anoints his head with expensive unguents in a public act that sparks an indignant response from bystanders and elicits from Jesus a passionate defense (14:3–9).
- Women are the witness to three final, crucial episodes: his torture and death on Golgatha (15: 40–41), his burial rituals (15:47), and an encounter with an angel at his empty tomb (16:1–8).[6]

Mark's first two anecdotes of Jesus and women follow patriarchal norms. Jesus cures Peter's mother-in-law of a fever: She is in her home; other than her relationship with Peter, she has no name. She responds by serving Jesus and his male disciples and meeting their needs. Jairus's daughter resides in her father's home; she has no name or identity of her own. Her father speaks for her; she has no independent voice.[7]

But in the other encounters in Mark's gospel, women are presented as distinct individuals, almost entirely without reference to the men in their lives. They encounter Jesus in public places rather than within the confines of home. These women take unusual initiative, instigating encounters with Jesus that defy cultural norms.

The woman suffering from a twelve-year hemorrhage interrupts and provides additional suspense in the story of the raising of Jairus's daughter. The episode stands out on one level because of Mark's clinical description of what is likely a gynecological issue. If the bleeding had come from somewhere other than the woman's vagina, it is likely that Mark would have named the source.[8] According to Leviticus 15:19–27, such bleeding makes women unclean. This woman's decision to seek Jesus out in a public place in a state of perceived uncleanness, reaching out to touch his clothing, is audacious. Will she render him unclean, as Leviticus decrees? She touches him, confident that her initiative will result in a cure. And it does. The woman is not given a name, but she is also not referred to in relation to a male.

The feisty Syrophoenician woman seeking a cure for a daughter possessed by a demon also takes an unconventional initiative: As a Gentile, she crosses a religious and ethnic boundary, seeking out the Judean teacher in public and falling at his feet. That is not the limit of her boldness. Jesus initially refuses her request, speaking in a proverb that implies that she along with all others who are not Israelites are dogs and that the healing for which she is asking is "bread" meant for "the children," presumably the children of Israel. Undeterred, she accepts his insult, turning it to her advantage with a cleverness that Jesus appreciates: Even dogs under the table deserve some scraps, she tells him. Instead of disapproving of the woman's pushiness, he commends her for her clever comeback and pronounces the daughter cured.

## JESUS AND DIVORCE

No actual female characters appear in Jesus's argument with the Pharisees bent on trapping him into a debate on the legality of divorce. His answers can be heard as a defense of marriage and as a pointed criticism of the way Mosaic law enables the short sexual attention span of some men.

> The Pharisees approached and asked, "Is it lawful for a husband to divorce his wife?" They were testing him. He said to them in reply, "What did Moses command you?" They replied, "Moses permitted him to write a bill of divorce and dismiss her." But Jesus told them, "Because of the hardness of your hearts he wrote you this commandment. But from the beginning of creation, 'God made them male and female. For this reason a man shall leave his father and mother [and be joined to his wife], and the two shall become one flesh.' So they are no longer two but one flesh. Therefore what God has joined together, no human being must separate." In the house the disciples again questioned him about this. He said to them, "Whoever divorces his wife and marries another commits adultery against her; and if she divorces her husband and marries another, she commits adultery." (Mark 10:2–12)

This encounter is more evidence of Jesus's intent to reverse the curse in Genesis that rendered women inferior to men. By echoing the Creator's intent to make them male and female and that the two should become "one flesh," Jesus's rebuke of the Pharisees is a reminder of the original equality and harmony that existed between man and woman before the Fall and the curse of male domination. It is a call to the church and the world in the present day to continue the work of restoring women to full equality with men.[9]

Jesus's opposition to divorce lands us in a hornet's nest of complexities. To make sense of it we must keep the big picture in mind: Jesus's inauguration of the reign of God that brings healing to many forms of brokenness. This prohibition of divorce is the only time Mark explicitly refers to the Genesis creation story. He invokes God's original intent to create male and female and have them experience union with one another. On that basis he exhorts men not to divorce their wives. Divorce not only violates God's will that husband and wife should be "one flesh." In the patriarchal culture of Jesus's day, divorce was primarily if not exclusively a male option, so it perpetuated the curse of male domination over women.[10] As exclusive breadwinner and owner of the family property, the husband was the dominant party in the marriage. If a man chose to divorce his wife, she had no recourse and no means of support.

Jesus's move to heal this form of brokenness differs from his other acts of healing. In place of actions that aim to re-create a world in which people are well fed, healthy, and in harmony with one another, this teaching appears to be clear-cut and doable. Prohibiting divorce may create a context for a couple to achieve the union of a man and a woman living under the restored reign of God. But the unbroken marriage too often perpetuates a relationship of male domination.

The challenge of maintaining union, whether between men and women or among mutual enemies, is an ongoing one, initiated by Jesus, taught to his followers by precept and example and taken up by them after his death and resurrection. We do not achieve it but can only taste it until God brings it to fullness at some unknown time. And the circumstances in which such union is pursued are constantly changing. The problem with Jesus's teaching prohibiting divorce is that it has been all too easy for the Church to reduce the challenge of brokenness to something that is so easily legislated and ignore the reasoning behind the prohibition. The Church has prohibited divorce—even legislated against it—and has left male domination of women in place.

Matthew and Paul were each aware of Jesus's teaching and repeated it, but they also made exceptions to the rule. Matthew modifies the teaching as Mark records it, introducing what has been known as "the exception clause": "unless the marriage is unlawful (*porneias*)" (NABRE) or "except on the ground of unchastity" (NRSV) (Matthew 5:32; similarly Matthew 19:9). In 1 Corinthians 7:10–11, Paul draws on essentially the same instruction of Jesus in Mark:

> To the married, however, I give this instruction (not I, but the Lord): A wife should not separate from her husband—and if she does separate she must either remain single or become reconciled to her husband—and a husband should not divorce his wife.

But then Paul goes on also to make an exception:

> [I]f any brother has a wife who is an unbeliever, and she is willing to go on living with him, he should not divorce her; and if any woman has a husband who is an unbeliever, and he is willing to go on living with her, she should not divorce her husband. . . . If the unbeliever separates, however, let him separate. The brother or sister is not bound in such cases; God has called you to peace. (vv 12–13, 15)

The reason Paul permits divorce in the case of an unbeliever not wishing to be married to a believer shares the spirit of Jesus's reason for prohibiting it: the goal to be served is in both cases harmony, peace, the healing of brokenness.[11]

The Church has always faced the challenge of responding to new situations that Jesus did not or could not have anticipated. Like Paul, we are obliged to take into account circumstances that work against the larger goal of healing in a broken world. Generally speaking, it is good for a couple to resolve their conflicts, strengthen their bond, and deepen their love. But what should a wife do when her spouse is abusive? Won't forcing a wife to remain in an abusive marriage only perpetuate brokenness? Obedience to Jesus's prohibition of divorce should not serve to perpetuate the curse of male domination.[12] We must not be deceived by thinking that Jesus's opposition to divorce is clear-cut and doable. It is far more difficult than simply obeying the rule. When we look no further than the rule to the goal the rule is designed to achieve, we may well subvert the goal and perpetuate the curse.

## SERVING AND FOLLOWING

Elizabeth Struthers Malbon has written perceptively about women in the Gospel of Mark. In *Mark's Jesus: Characterizations as Narrative Christology*[13] she looks at two encounters that place the spotlight on women: the widow who gives all of her meager savings to the Temple (12:41–44) and the unnamed woman in Bethany who, in an extravagant display of veneration, anoints the head of Jesus (14:3–9).

Mark sets the scene for the widow's story in the last days of Jesus's public ministry, when he was teaching in the Temple of Jerusalem. He warns his listeners about the hypocrisies of opulently robed scribes receiving accolades in the marketplace, grabbing the best seats at the synagogue and places of honor at banquet tables. He points out that these first-century one-percenters are made rich with ill-gotten gains: "They devour the houses of widows and, as a pretext, recite lengthy prayers. They will receive a very severe condemnation" (Mark 12:40).

Then the scene shifts to depict Jesus, again in the Temple, but now seated in view of the Temple treasury, where a parade of worshippers, rich and poor, can be seen throwing money into the box. Mark observes that while the affluent make ostentatious offerings, it is the poor widow's offering of two copper coins that earns praise. "Amen, I say to you, this poor widow put in more than all the other contributors to the treasury. For they have all contributed from their surplus wealth, but she, from her poverty, has contributed all she had, her whole livelihood"—literally translated, "her whole life."

In the second story, Jesus is again sitting, but this time in the house of Simon the Leper. An anonymous woman enters with a jar of ointment, which she proceeds to pour on Jesus's head. Her gift is extraordinarily costly, three hundred denarii. For an Israelite laborer who typically received one denarius for a day's work, if he rested on the Sabbath, this would be close to a year's earnings.

Malbon points out that it is a deliberate act of provocative teaching for Mark to juxtapose with uncomplimentary stories about men both the story of the widow who puts her whole life into the treasury of God's Temple and that of the woman who anoints Jesus with an extravagantly expensive ointment. The scribes who "devour the houses of widows" instead of caring for the vulnerable violate the responsibility decreed for them in the scriptures. The veneration by the woman in Bethany is bracketed by two starkly contrasting scenes of malice: the plot by Temple authorities to have Jesus quietly arrested so as not to anger his supporters and Judas's decision to accept payment from them in exchange for delivering him into their hands.

Malbon proposes that in contrast to the abusive scribes and the scheming Judas and Temple authorities, the widow who gives "her whole life" and the woman who anoints Jesus's head with an ointment costing a year's wages are both "exemplars of Jesus' self-giving service," which Jesus names in Mark 10:45 as the hallmark of how to be one of his followers. To understand the distinction in Mark's gospel between the concept of servitude (the subordination of one person to another) and Jesus's transformational idea of service, consider how Jesus's power-hungry male disciples struggle to grasp the concept.[14]

Twice during their journey to Jerusalem, the disciples express their aspirations for greatness. In Mark 9 they argue about who is greater among them. In Mark 10 James and John express their desire to be given the places of honor in Jesus's kingdom. Both times Jesus responds by telling them that the one who is great is the one who takes the role of a "servant" (*diakonos*). In 10:45 he presents himself as their model, declaring that "the Son of Man did not come to be served (*diakonethenai*, a Greek passive verb) but to serve (*diakonesai*, a Greek active verb) and to give his life as a ransom for many." In other words, Jesus names service, *diakonia* in Greek, as the central task of his mission. This is deeply ironic, because service, in Israelite culture, is the work of women.[15]

The women who Mark in 15:41 belatedly informs us "followed" Jesus during his public life appear to have played the very role to which Jesus called his male disciples. They are described as having also "served" (*diekonoun*) him, the very word he uses to describe the essence of his mission to the Earth and the source of greatness for his male disciples.[16]

The women who "followed" and "served" Jesus are therefore the ones who stand by Jesus in his hour of greatest need while the male disciples, whom he called to "follow" him and to devote themselves to "service" as he himself has been doing, have abandoned him. This is another of the treasures embedded in Mark's narrative.

Horsley observes that in Mark's unfolding plotline, the importance of women increases as the loyalty of male disciples goes downhill. By the end of Mark's story, Jesus arrives in Jerusalem to meet his fate in the company of many women, who had both been "following him" and "serving him" (Mark 15:40–41, authors' translation). Those are two important roles. To "follow" Jesus implies the women are on equal footing with the male disciples. At the beginning of Mark's narrative, Peter and Andrew, James and John, and later Levi enthusiastically leave everything and follow Jesus. To "serve" Jesus, as we shall see, implies much more.

By the time Jesus gets to Golgotha, the place of his execution, the male disciples have all abandoned him. At this point, Mark introduces three female characters by their own names: Magdalene, Salome, and Mary, the mother of James the younger and Joses. Belatedly he tells us that these and many other women have followed and served Jesus while he was in Galilee (15:40–41). Only the two Marys and Salome remain with Jesus while he suffers and dies. After his death, the two Marys watch as he is being buried (15:47). When the Sabbath has passed the three visit his tomb to perform the final, intimate service that the living do for the dead in Israelite culture: washing his body and anointing it for burial (16:1).

Mark ends his story with a shock. When the women arrive at the tomb to anoint Jesus's body, they find it gone. Instead of the body the tomb is occupied by a young man clothed in white, an angelic presence who declares that Jesus has risen. He invites the women to look around and see for themselves that the body is not there. The angel then instructs them to find Peter and the other terrorized disciples and tell them to go to Galilee, where Jesus will meet them in faithfulness to the promise he made on the way the Mount of Olives before he was arrested. Instead of obeying the angel, they panic and flee, just as the male disciples had done when Jesus was arrested; and to no one do they say a word.

Christians struggled with this ending from earliest times. Many modern translations inform Bible readers that manuscripts from the first few centuries after Jesus end the gospel in a variety of ways. Most critical scholars judge the manuscripts that end at 16:8 to be the most reliable. But this is a very disturbing ending. For one thing, it clearly conflicts not only with the endings of the other three gospels but with what for the believers was historical fact: someone must have spread the news about the empty tomb; someone encountered Jesus alive. Otherwise there would be no Christian Church. Three other endings in the manuscripts seem designed to address this problem. The most popular of these scholars refer to as "The Longer Ending." It narrates a series of three appearances of Jesus to which various followers respond with disbelief. The third time Jesus takes the disciples to task for not believing, and commissions them with the words, "Go into the whole world and proclaim the gospel to every creature." This ending was the most satisfactory one for Christians, became the authoritative ending for the Church, was copied most often, and consequently was numbered as Mark 16:9–20. But it appears to be pieced together from the other three gospels and is not Mark's original ending.[17]

Mark's original ending completes the pattern of women in his Gospel emerging from the restrictions of patriarchal culture into being individuals with their own names, who act on behalf of their own needs or with the gifts they have to offer.

Why does Mark end his story with an image of once-faithful and courageous women fleeing the empty tomb, terrified to speak of Jesus's resurrection?

Thomas E. Boomershine, an expert in oral storytelling, recognized long ago that Mark's gospel is not historical but rhetorical; it is a work of narrative preaching.[18] Mark has an urgent message to people who believe in Jesus but are cowering in fear of the consequences of proclaiming the resurrection and pursuing a life of healing and reconciliation that flows from the freely chosen and joyful embrace of God's rule. By ending his story with the failure of even these faithful women to tear off and dispatch the frightened disciples to Galilee where they will meet the living Jesus, Mark poses a difficult question: Who will tell the world the story of the resurrection of Jesus?

To those who first heard Mark's story, the message of the resurrection involved a new way of life that would require courage because it disrupted social and economic and political structures, including male domination of women. Will successive generations of listeners have the courage to accomplish what the terrified women who fled the empty tomb were unable to do?

In Mark's story there are no longer male heroes for the women to depend on. They have all fled. Nor does Jesus treat women as people whose proper place is under the supervision of their men. By the end of Mark's narrative both men and women are "followers" and "servants" with the freedom and responsibility for taking initiative in public for living life free of the curse. It is a daunting challenge. The social structures of this world will resist. If we are to return to a life of harmony with God that will change the world, we will need to work together, looking to the Holy Spirit to equip us (Mark 13:11). The human species has often exhibited the capacity to say "no" to what is, and "yes" to what now seems impossible.

For Mark's Israelite listeners, male violence and authority during the Great Revolt of 66–70 CE has resulted in disaster. Striving to be great by lording it over people and exercising authority over them is not the way of the Son of Man (Mark 10:42–45). Service is at the heart of his mission, but service is not servility. It is exercised by both men and women by taking initiative in the public world, a free initiative that the Creator granted to males and females equally. They are made in the image of their Creator, male and female equally. They are given dominion over creation, not to exploit it in the manner of the great and powerful rulers in this fallen world, but to give their lives to ransom this world so that it may be free.

## A RADICAL REVISION

We cannot end this exposition of the motif of "service" in Jesus's mission and the mission of those who would follow him without addressing the concern expressed by many women today about being relegated to a secondary role to men in the ministry of the church and in society.

Women have rightly pointed out that what they need to hear today is not a call to serve, but to break out of that culturally imposed role. A "gospel" that calls women to service, but not to leadership or autonomy, is not "good news." It is a call to remain in a state of oppression. It is *men* who need to hear this message, not women. Women need to hear a call that somehow gives them power over their own lives.[19]

One of the great ironies of Mark's gospel is not only that the narrator suggests the traditionally female role of service is a core characteristic of Jesus's mission. He also portrays women as breaking out of the cultural boundaries of Jesus's society. Thus Mark cracks open the whole issue of the relationship between men and women in a world whose ruling system is being challenged and transformed. Jesus summons all people, male and female, to embrace a revolutionary set of values. Neither men nor women are restricted by patriarchal norms as they seek creative ways to serve and restore our lost harmony with each other and all creation.

Insights into Jesus's path-breaking encounters with women in the Gospel of Mark are useful to today's listening or reading audience. Although we also live in a largely patriarchal culture, the specific manifestations of patriarchy in Jesus's culture are not the same as in ours. A woman taking initiative in addressing in public a male other than a family member is so commonplace in Western culture that contemporary readers of the gospel would not see a shocking violation of patriarchal norms. This is one reason readers and listeners today often fail to recognize the significance of how the women in Mark behave and how Jesus responds.

A second reason is that Christians rarely hear or read Mark's gospel as a unified narrative. The standard Sunday-morning worship service in most traditional Christian dominations has largely been shaped by the practice of lectionary reading. The gospel stories and the rest of the Bible have been deconstructed into a series of isolated vignettes. Listeners still hear only a single or very small group of the episodes of Mark's narrative at any one time. Few worshippers are given the opportunity to hear the entire narrative from beginning to end. But the impact of any given story about women in Mark is best perceived in the context of the entire narrative.

## THE ROOTS OF JESUS'S LIBERATING WORK

The female characters in Mark's narrative are not the first Israelite women to break the pattern of behavior set for them in a patriarchal culture. Carol Meyers's survey of women in the Hebrew scriptures in *The New Interpreter's Dictionary of the Bible*[20] confirms the patriarchal character of Israelite society and the conformity of most women mentioned in those scriptures to essential roles in the home as well as the practice of naming them as some man's wife, mother, or daughter. But there are notable exceptions.

The nurse for the matriarch Rebekah is simply named Deborah (Genesis 35:8). The two midwives who disobeyed Pharaoh's order to kill male newborns are named Shiphrah and Puah (Exodus 1:15). The harlot of Jericho, Rahab, is understandably enough named without association with a male. And while most translations designate the judge Deborah as "wife of Lappidoth," Meyers points out that Lappidoth could be the name of her hometown rather than her husband. In that case it would be translated "a woman of [the town of] Lappidoth" (Judges 4:4). Besides being named independently of a male, Miriam (Exodus 15; Numbers 12) and Deborah (Judges 4–5) clearly play a public role. Miriam (Exodus 15:20), Deborah (Judges 4:4), Huldah (2 Chronicles 34:22), and Noadiah (Nehemiah 6:14) are called prophets. We may also note that three of the four women named in Matthew's genealogy of Jesus, Tamar (Genesis 38), Rahab (Joshua 2), and Ruth, all take initiative in relation to men.

It is not only Israel's earlier scriptures that provide precedents and inspiration for women to surmount the obstacles to their participation in public life. In the century just prior to Jesus's ministry a few exceptional Gentile women had held political power in the Roman Empire, most famously Cleopatra VII in Egypt. In the wider Gentile world in which Mark probably told his story of Jesus, Gentile widows of male citizens were able to inherit property and a few women became wealthy from business enterprises.[21] Documentary sources like dedicatory inscriptions, funerary epitaphs, statues, and portraits highlight the importance of individual women. In Israelite communities across the empire, women held offices like "ruler of the synagogue," "mother of the synagogue," "elder," and "priest."[22] Mark's own narrative witnesses to the presence in Galilee of women with the means and independence to be among Jesus's followers and to help support him and those who accompanied him as he went around preaching and healing (Mark 15:40–41). Clearly, the women who venture into public space to encounter Jesus do so on their own initiative. Jesus does not proclaim their freedom to do so or invite them; he welcomes them when they take initiative.[23]

The picture of Jesus that emerges from this information about Jesus's world is less that of an innovator than of a leader who drew on conflicted traditions and marginal impulses in his culture to muster a new movement that would release God's healing power into a broken world. Consequently, while Jesus welcomes women who violate the norms of his patriarchal culture, their taking the initiative to do so did not originate with Jesus but was one of the many currents of innovation present in the culture of Israel.

## CONFLICTING VIEWS OF WOMEN IN PAUL'S LETTERS

Mark's gospel clearly shows that Jesus treated women as equal partners in his work of transforming the world. But how are we to reconcile this idea with longstanding tradition in Christianity, often authorized by appeals to the writings of St. Paul, that women should be subservient to men?

Saul of Tarsus was an Israelite who lived in a lively multicultural city of the Diaspora.[24] He was not among the original twelve apostles, but according to the autobiographical account in his letter to the Galatians, he had a life-changing experience when Jesus called him to preach the gospel to Gentiles across the Roman Empire. Israelites living in the Gentile world often had both a Semitic and Greco-Roman name. Beginning in Acts 13, Saul became known as Paul, a Greco-Roman name.[25]

In Paul's letters to the various churches he founded or worked with, he is revealed as a complex thinker. He is not always consistent; his views grow and change, as befits a man of faith who is still working things out.

Statements about women in letters to nascent Christian communities attributed to Paul are all over the map: In some, Paul praises the ministry of women and holds them up as an example to all. In others, Paul supports the same biases against women that prevailed in Israelite and Gentile culture. The challenge for scripture scholars and ordinary believers who try to sort out his disparate views is to identify the most accurate and authentic sources for the words attributed to Paul.

Seven of the thirteen letters that bear Paul's name—Romans, 1 Corinthians, 2 Corinthians, Galatians, Philippians, 1 Thessalonians, and Philemon—are generally accepted as authentic. The others, according to many scripture scholars, are composed by his successors who attribute their compositions to Paul in the belief that they are faithfully representing him. In Paul's own time, his original words may have been changed by editors who prepared his letters for wider distribution. Years later, other manuscripts may have been changed by church leaders who inserted their biases into Paul's writings.

It was a common practice in the ancient world for a prominent public figure to collect and publish letters. Writing letters was an occasion to articulate one's views on matters of public concern. In the ancient world, gathering the letters of an influential figure into one convenient collection was a way to make a person's views available to a wider public. Roman orator Marcus Tullius Cicero instructed a man named Atticus to see that his letters were collected and published.[26]

Here is an account, written about 200 CE by the Christian theologian Tertullian, of a dispute over the credibility of a letter attributed to Paul:

> But if they claim writings which are wrongly inscribed with Paul's name—I mean the example of Thecla—in support of women's freedom to teach and baptize, let them know that a presbyter in Asia, who put together that book, heaping up a narrative as it were from his own materials under Paul's name, when after conviction he confessed that he had done it from the love of Paul, resigned his position. (*de Baptismo* 1.17)[27]

In the letters that all scholars agree are by Paul himself there are numerous passages about women in marriage and women in the church in which women are viewed as equal to men. Women are frequently named as partners in leadership of the church with Paul without any hint that they are under his authority. A woman named Junia is named as an "apostle" alongside her husband Andronicus in Romans 16:7. Phoebe is named as a "deacon" in Romans 16:1. According to 1 Corinthians 11:4–5 both men and women prophesy in the assembly. In 1 Corinthians 14:26–33 Paul provides a picture of what took place in Christian assemblies. From verses 29–32, it sounds as though various members "prophesied" by sharing revelations that would teach and encourage the community. Whether one person's revelation was really from the Lord or not had to be evaluated in some way by others. According to 1 Corinthians 11:4–5 both men and women prophesied. The men do not cover their heads, while the women do, but both speak. And yet in 1 Corinthians 11:3, Paul states clearly that the man is head of the woman.

In his 1994 book, *Paul: The Apostle to America*,[28] Robert Jewett proposes that over the course of his life Paul, in response to the real equality women enjoyed in the early church, was growing out of a traditional attitude toward women that he learned growing up in the prevailing patriarchal society. But because over the course of time Paul was not consistent in his views, socially conservative successors, like the authors of letters to Timothy, Titus, Colossians, and perhaps Ephesians, found support in his letters for their position, while socially progressive successors, like the author of *The Acts of Paul and Thecla*, found support for theirs.[29] Clearly those who collected the letters of Paul that are included in the New Testament were drawn to the letters that express a socially conservative view of women. For almost the entire history of the Church, the statements that clearly prescribe subordination of women in letters like Ephesians, Colossians, and 1 Timothy[30] have blinded readers to the multitude of passages in Paul's genuine letters that give a contrasting picture. Letters by socially conservative successors are uncritically attributed to Paul. The quotation from Tertullian above shows that socially conservative Christians did not hesitate to view as forgeries works in the Pauline tradition in which the Apostle was instrumental in liberating women from patriarchal restrictions.

## THE CRUCIAL TEXT

The most frequently quoted passage favoring women's subordination in the Church is 1 Corinthians 14:33b–36, which, of course, is found in what all agree is a letter by Paul himself:

> Let the women be silent in the assemblies. For it is not permitted for them to speak, but let them be in submission, as the law indeed says. And if they wish to

> learn, let them ask their husbands at home; for it is disgraceful for a woman to speak in the assembly. Is it from you that the word of God came forth? Or is it to you alone that it has come? (Adam Bartholomew provisional translation)

This directive appears in the middle of what Paul has to say about the practice of prophesying in the Church's gatherings. It conflicts with 1 Corinthians 11:4–5, which shows that women clearly prophesy in the assembly. Many scholars also find that 1 Corinthians 14:33–36 disrupts the flow of what Paul says before and after. These observations have led many scholars to conclude that what is now 1 Corinthians 14:33–36 was inserted into Paul's letter by a later socially conservative editor.

A more plausible explanation proposed by Charles Talbert in *Reading Corinthians* (91–93) is that Paul is quoting here the words of people with whom he disagrees, as he does elsewhere in 1 Corinthians.[31] The writing system for Greek in Paul's day did not include quotation marks to identify the words of an opposing view. The clearest indication that Talbert is right is the gender of the word "alone" in Paul's rhetorical question, "Or is it to you alone that it has come?" The word translated "alone" is *monous*. This is a masculine form of the adjective that could be generic, referring to women together with men. So he must be rebutting the view that the word of God comes only to males: "Is it from you that the word of God came forth? Or is it to you [males] alone that it has come?" If he were taking the women to task, asking, "Is it from you that the word of God came forth? Or is it to you [females] alone that it has come?" he would have to have used the feminine form *monas*. The words that precede Paul's rhetorical questions should be bracketed with quotation marks to show that they express the view of his opponents:

> "Let the women be silent in the assemblies. For it is not permitted for them to speak, but let them be in submission, as the law indeed says. And if they wish to learn, let them ask their husbands at home; for it is disgraceful for a woman to speak in the assembly." Is it from you that the word of God came forth? Or is it to you males alone that it has come? (Adam Bartholomew revised translation)

This passage from 1 Corinthians has played a prominent role in the silencing of women's voices in the Church. It is stunning to observe how a seemingly small thing like paying attention to the gender of one little but crucial Greek word, instead of reinforcing the curse of male domination could have promoted Jesus's mission of lifting that curse, in the Church and in society as a whole.

## WILL THE REAL PAUL PLEASE STAND UP?

The letters of Paul in no way consistently counsel the subordination and silencing of women. The complexity of the process by which the collecting of Paul's letters would have come about, together with the explicit evidence that his successors represented only one side of his complex thinking, shows that the question of the place of women in the Church was one about which there were various points of view.

According to Mark's revolutionary narrative, Jesus's actions aimed at restoring the world order that preceded the fall of Adam and Eve. And there is plenty of evidence in both scripture and sources not included in the orthodox canon that women played leadership roles in the early church.[32] But rather than sustaining the liberation that Jesus had begun, the emerging Church leaders caved in to a worldview that viewed women as inferior to men. Many Christians today are now recognizing that the dominant tradition of the Church has failed to be faithful to Jesus's mission of freeing women from male domination.

Since the Protestant Reformation, the idea of ordaining women has been acceptable to a number of denominations. Some Reformed, Presbyterian, Methodist, Baptist, historically black denominations, and independent churches began the practice without contest. Churches in the worldwide Anglican Communion struggle with this issue as a result of deep cultural differences. The Episcopal Church in the United States, the Anglican Church of Canada, the Church of England, and other member churches began ordaining women as deacons, priests, and bishops in the late twentieth century.

In 1976 the Sacred Congregation for the Doctrine of the Faith of the Roman Catholic Church addressed for the first time the issue of women's ordination. This was partly in response to the ordination of eleven Episcopal women as priests in Philadelphia in 1974. The result was the document "Declaration on the Question of the Admission of Women to the Ministerial Priesthood." The Vatican's conclusion was unequivocal and its traditional stance remained unchanged: Women cannot be ordained priests.[33] However, the issue did not go away, and in 1994 Pope John Paul II issued an Apostolic Letter, *Ordinatio Sacerdotalis*, that reiterated that the faithful were to hold as definitive the view that the Church had no authority to ordain women as priests.

In none of this discussion was the question of women deacons raised. In 2016 Pope Francis moved to establish a commission to study the possibility of women being ordained as deacons, a question not addressed in the 1976 "Declaration." One member of the commission is Phyllis Zagano, a prominent US Catholic scholar and author of *Holy Saturday: An Argument for the Restoration of the Female Diaconate.*[34] She notes that the Church ordained women from the earliest centuries until the Middle Ages, when the diaconate faded away as a separate order and was converted into a step toward priesthood. Zagano acknowledges that ordaining women as priests is simply not the teaching of the Catholic Church. But that doesn't mean women have no ordained role. In an interview for *U.S. Catholic* she commented:

> To say that a woman is not ordainable and cannot serve *in persona Christi*—as a deacon, in the person of Christ the servant—is to argue against the incarnation. The important thing is not that Christ became male. It's that Christ became human. If we say that a woman cannot live *in persona Christi*, I think we're making a terribly negative comment about the female gender.[35]

Sociological studies of the Christian movement in the first five centuries support Zagano's claims. They show that the ratio of women to men was disproportionately high, a situation that typically leads to greater freedom and power in a community. The evidence for women deacons in the early Church includes not only Paul's attributing that title to Phoebe but the prescription for the character of female deacons in 1 Timothy and a letter from the Roman governor Pliny in which he speaks of two Christian women with the title of "deacon."[36]

Where this will lead remains to be seen. What is clear is that even the age-old tradition of limiting ordination to males in the Roman Catholic Church requires defense and continues to be discussed.

One of the messages implied in Mark's narrative is that women, who exemplify "service," the core characteristic of Jesus's mission, can break out of the straitjacket of the social rules and exploitative relationships that a male-dominated society has imposed on them. While the women in Mark's gospel perform service to others in an exemplary way, no longer are they named only in relation to men; they do not serve under the direction of men; they are not dominated by them; they are not passive in their presence. They take initiative. This tension between traditional roles and freedom from norms imposed by males raises the question: How might women serve when they are free to explore their own ways of doing it?

Mark's Jesus calls men to learn how to relate to women in a way free of domination. He also calls men to learn from women how to be servants themselves. Not only are men asked to stop dominating women; Jesus asks men to sit at women's feet and learn from them.

# 5

# Violence and Death

In the foundational narrative of Genesis, death was not part of the original plan. Mortality comes as a consequence of Adam and Eve's decision to embrace evil as well as good. But unavoidable death is just the beginning of the downward spiral. Violence enters the storyline with their offspring, when the jealous Cain lures his more favored brother Abel to a deserted field. There, in a bleak act of jealous rage, he takes Abel's life, and earns an additional curse that renders him a perennial fugitive, banished from the once-fruitful Earth to a lifetime in the wilderness. Later on in the story, Cain's descendant Lamech introduces vengeance to the mix, bragging that his capacity for violence far surpasses that of his forebear, exulting in his appetite for revenge:

> I have killed a man for wounding me,
> a young man for bruising me.
> If Cain is avenged seven times,
> then Lamech seventy-seven times. (Genesis 4:23b–24)

As agents of death, the characters of Cain and Lamech accelerate a sharp turn in the Creation story, away from the lush Paradise of the original garden, and farther into the harsh conditions of a barren wilderness. These two are emblems of the hatreds between individuals, tribes, factions, and nations that have shaped human history.

## CHAPTER 5

The primeval myth of human origins concludes in the eleventh chapter of Genesis, the story of Noah's descendants, an autonomous and ambitious migrant people. The narrator explains that this is a time when the whole Earth is united by one language; all people speak and understand the same words. As Noah's descendants wander westward through the post-Paradise landscape, they find a promising spot to settle on the plains of a land called Shinar. There they resolve to manufacture the bricks that would build a great city; they begin to erect a proud tower, planting the seeds of a civilization that would extend its ambitions all the way to the heavens. Their clever leveraging of free will and a knack for technological innovation are immediately stymied by a Creator distressed by human arrogance. God strikes back at the uncompleted tower, using words as a weapon, splintering their common language into multiple dialects. Thus the Creator makes the members of this once unified people strangers to one another, and disperses them across the face of the Earth. As they depart their unfinished city, Noah's offspring, now bereft of a common language, name it Babel, a place of incoherence, fragmentation, and discord.

The Creator's strategy of sowing confusion with words can be interpreted on both a literal and spiritual level. Languages and dialects can be mutually unintelligible even when they are closely related, and people who truly desire to communicate can overcome any linguistic barrier. Yet the human spirit can also set itself against mutual intelligibility, even among people who share the same vocabulary and grammar.

The same relentless drive for power and profit that drew the Creator's sanction in the Genesis narrative has driven humanity down through the ages. Heedless humans have treated nature as a platform for their ambitions, exploiting and enslaving one other, waging wars, and decimating the fertile Earth to parched and unproductive land. The result has been havoc in both the natural world and the human community.

The Genesis narrative moves on from the wanderings of the linguistically divided descendants of Noah to a key moment in the story of Israel. God calls Abraham, a childless septuagenarian comfortably settled with his wife Sarah in Ur of the Chaldees, to establish a covenant. They agree to abandon their established home and migrate to Canaan, a land they do not know, in exchange for a promise of new life, land, and an abundance of descendants. God's command and his promise to the couple are complex. Their name will be great and their lineage will be an instrument through which "all the families of the earth will find blessing" (Genesis 12:1–3). This marks the beginning of the story of Israel, and the promise that this nation, twelve tribes of Abraham's and Sarah's descendants, will be the key to the rest of the world's story. It is a narrative that remained central to Israel's self-consciousness throughout its history.

A number of Israel's prophets imagine the consummation of this great calling. The vision of Isaiah (2:2–4) could hardly be more majestic:

> In days to come,
> The mountain of the Lord's house
> shall be established as the highest mountain
> and raised above the hills.
> All nations shall stream toward it.
> Many peoples shall come and say:
> "Come, let us go up to the Lord's mountain,
> to the house of the God of Jacob,
> That he may instruct us in his ways,
> and we may walk in his paths."
> For from Zion shall go forth instruction,
> and the word of the Lord from Jerusalem.
> He shall judge between the nations,
> and set terms for many peoples.
> They shall beat their swords into plowshares
> and their spears into pruning hooks;
> One nation shall not raise the sword against another,
> nor shall they train for war again.

The prophet Micah repeats Isaiah's vision almost word for word (Micah 4:1–4) and Zechariah echoes it (8:20–23). The Isaiah who articulated this vision in Isaiah 2 was the first of three prophets whose oracles are included in our present book of Isaiah. The third Isaiah captured the vision of his predecessor more briefly in words from which Jesus would later quote (Isaiah 56:7):

Them I will bring to my holy mountain
and make them joyful in my house of prayer;
Their burnt offerings and their sacrifices
will be acceptable on my altar,
For my house shall be called
a house of prayer for all peoples.

Centuries later the prophet Daniel (7:13–14) experiences a similar vision. In his night vision he saw all the nations, peoples, and tongues of the world drawn together under the humane rule of "One like a son of man," then identified as the "people of the holy ones of the Most High" (7:27), the people of Israel. Jesus envisions his own ministry of reconciliation brought to fulfillment when on behalf of Israel, at some unknown day and hour after his death and resurrection, he will be seen "coming with the clouds" (Daniel 7:13; Mark 14:62).[1]

## HEALING HUMAN DIVISIONS

In Mark's narrative Jesus speaks of himself as the "Son of Man," referring to his humanity, not his maleness, and better translated "the human one."[2] The phrase is a clear allusion to Daniel's vision for all the nations of the world. Jesus devotes his ministry to healing human divisions from almost the very beginning. He first addresses the social and political divisions within Israel itself. In Jesus's version of a new covenant, no Israelite would be excluded from the rule of God he is initiating—not sinners, not even the loathed tax collectors. His choice of twelve apostles to join him in his work symbolized in a powerful way that the twelve tribes of Israel would be gathered back from the dispersion that resulted from successive conquests, first of the northern ten tribes by the Assyrians in 722 BCE and then of Judah and Benjamin by the Babylonians in 587 BCE.[3]

Jesus includes in the circle of his closest associates both Simon the Cananean, a freedom fighter,[4] and Levi the tax collector.[5] Their politics are polar opposites. People like Simon advocate violent revolt against Rome; tax collectors like Levi cooperate with the Romans and enrich themselves and the already wealthy Romans. Mark's story of the call of Levi (2:13–14) echoes the stories of the calling of Peter and Andrew and of James and John (1:16–20). Apparently, for Jesus, Israel could be purified of disobedience to God's rule only through an inclusion that would overcome the ongoing discord among the various factions of the people of Israel.

But Jesus does not gather the tribes of Israel back together strictly for their own sake. God had called the nation into being so the descendants of Abraham and Sarah might be a blessing to all the families of the Earth (Genesis 12:3). As long as Israel is scattered in exile it cannot fulfill its destiny. The prophet Ezekiel interpreted the scattering of Israel among the nations as the result of Israel's disobedience to God and an ongoing offense against God's name among the nations. He envisioned God's regathering of the scattered people back as a means of honoring God's name among the nations once again (Ezekiel 36).

In fulfillment of the call of Abraham and Sarah and the visions of Israel's prophets, Jesus also includes the Gentiles in his compassionate work. It begins with the healing of a man possessed by a legion of demons in the Gentile region of Gerasa (5:1–20). Shortly thereafter, Jesus challenges the Pharisees' devotion to purification rituals that set Israelites apart and calls instead for a purity of heart that reaches far beyond the faith and rituals of Israel to include all people (7:1–23).[6] His encounter in Tyre with the Syrophoenician woman challenges him to heal her daughter of an unclean spirit just as he has been driving unclean spirits out of Israelites. Jesus's compassionate response to her pleas further transcends established boundaries of ethnicity and gender.[7] To underscore his mission to heal divisions of language and culture, Jesus then travels to the ten Gentile cities of the Decapolis. There he restores hearing and speech to a man who is deaf and mute and feeds four thousand Gentiles[8] with an abundant feast of bread and fish, just as earlier in Mark's narrative he fed five thousand Israelites.[9] This storyline reaches its climax when Jesus echoes the words of Isaiah that the Temple is to become a "house of prayer for all peoples" (Mark 11:17; Isaiah 56:7).[10] The phrase "all peoples" sets up echoes in Jesus's vision in Mark 13 that his followers will carry the gospel to all nations. In the story of the woman who anoints his body beforehand for burial, he declares that her deed will be remembered as the gospel is preached to the whole world. Implicit in Jesus's vision of the ultimate coming of the Son of Man is the establishment through him of God's humane rule over all nations.

CHAPTER 5

# CONQUERORS AND THE CONQUERED

The world Jesus lived in was full of military conquests, with Israelites often but not always on the losing side. In that world, it was common for victors to be welcomed into a city with a joyful ceremony that climaxed in a visit to the city's sanctuary, where the conqueror offered sacrifice to the gods or purified it of uncleanness.

Flavius Josephus, a contemporary of Jesus and Mark, tells the story of the high priest Jaddua, who, around 332 BCE, opened the gates of Jerusalem to Alexander the Great and his army after they had conquered Gaza. Jaddua had angered Alexander and was agonizing over what to do as the conqueror approached the city. In a dream, God assures him of divine protection. Jaddua, the priests, and a throng of citizens went out in procession to meet Alexander, who himself was guided by a dream and prostrated himself before Jaddua, who was clad in a headdress on which was engraved the name of God. Alexander then entered the city, went into the Temple, and offered sacrifice.[11]

Israel's own Greek scriptures included a vivid account of the Maccabean revolt: In 164 BCE, 150 years after Alexander established Greek rule in Israel, the hero Judas Maccabeus defeated the Greek vice-regent Lyceas and celebrated his victory with a similar ceremony. He entered Jerusalem with his army, then went immediately to the sanctuary, which he purified, tearing down the altar that had been defiled by the Gentiles (1 Maccabees 4:36–60). But the Israelite conquest of Jerusalem did not last. About twenty years later Judas's brother Simon recaptured the citadel in Jerusalem. "He expelled [the inhabitants] from the citadel and cleansed it of impurities. On the twenty-third day of the second month, in the one hundred and seventy-first year, the Jews entered the citadel with shouts of praise, the waving of palm branches, the playing of harps and cymbals and lyres, and the singing of hymns and canticles, because a great enemy of Israel had been crushed" (1 Maccabees 13:50–51). These and other stories of Israel's military heroes inspired yet another revolt, against Rome, in Jesus's time.

Jesus leverages the traditions of triumphal conquest by engaging in what Adelle Yarbro Collins depicts as a dramatic act of "street theater."[12] In Mark's narrative, he enters Jerusalem and cleanses its temple in a manner that evokes the ceremonies repeated many times by the conquering armies of Egypt, Greece, and Rome. But he is engaging in parody. His purpose is to drain these conquering ceremonies of their violence.

Mark's oral history of the life of Jesus was being told to Jewish audiences around 66–70 CE, during or in the wake of the Israelites' failed Great Revolt against Rome, and its imagery would resonate with a dispirited people who yearned to be free of the latest in a series of imperial oppressors. Jesus enters Jerusalem mounted on an animal Mark describes as a *polon*, the Greek word for a horse of any age. By itself the word is ambiguous. But in the context of the narrative of Jesus's dramatic entry without an army, it would readily remind Greek-speaking Israelites of the prophecy in Zechariah 9:9–10. The Greek translation of that prophecy pictures the claimant to the throne riding a donkey and a young *polon*.[13] Zechariah himself presents a twist on the typical scene of military conquest. While a *polon* could be a warhorse, a donkey is clearly not.[14] In Zechariah's vision this king will rid the world of the instruments of war and establish universal peace among the nations. Here is the scene in *A New English Translation of the Septuagint*:

> Rejoice greatly, O daughter Sion!
>    Proclaim, O daughter Ierousalem!
> Behold, your king comes to you,
>    just and salvific is he,
> meek and riding on a beast of burden and a young foal [*polon*].
> And he will utterly destroy chariots from Ephraim
>    and cavalry from Ierousalem,
> and the battle bow shall be destroyed,
>    and there shall be abundance and peace from nations,
> and he shall reign over the waters as far as the sea,
>    and the rivers at the exits of the earth.

Collins proposes that this bit of street theater was Mark's way of "giving intentionally conflicting messianic signals in order to provoke the readers to reconsider their presuppositions about messiahship, discipleship and the kingdom of God."[15] Christians often say that Jesus rejected Israel's longing for political liberation, that he was calling Israel to seek spiritual liberation instead. But Jesus was dedicated to the political liberation of Israel. His theatrical performance symbolically dramatizes his different strategy for achieving it that involved and had consequences through Israel for the whole world. He models for Israel freedom from fear of enemies, both fellow Israelites and Gentiles, in the face of imagined or even real violence. His path is a new spiritual path, but it remains focused on his people's political goal as their service to the nations. He pursues political liberation by seeking reconciliation with Israel's enemies where he can achieve it, even at the cost of his life. He teaches that path and after his resurrection is present to his followers, inspiring his way of fearless efforts at reconciliation in them. Instead of risking being killed in making war, Jesus and his followers will risk being killed in making peace.

Jesus's parody of military conquest does not end with his ceremonial entry into the city. Following the model of both Israelite and Gentile conquerors, Jesus also visits the Temple.[16] Like Judas Maccabeus and his brother Simon, he takes dramatic action to cleanse it. But in contrast to Judas and Simon, Jesus does not tear down an altar defiled by Gentile sacrifices. He does not cleanse the Temple by expelling the foreigners. That practice has pervaded human history to this day, climaxing in campaigns of "ethnic cleansing," the slaughter of those not of one's own tribe. Jesus cleanses the Temple by clearing a place for all peoples to join Israel in prayer.[17] Mark's narrative clearly shows Jesus's intent to cleanse the world of the violent conflict that pits nation against nation and perpetuates the curse of death.

Jesus was not alone in welcoming the Gentiles to join Israel in worship in the Temple or in urging peaceful ways of resolving conflicts between Israelites and their Roman rulers. In 26 CE when Pilate brought into Jerusalem military standards that the Israelites regarded as idolatrous, thousands of them staged a nonviolent protest for five days at the house of Pilate in Caesarea. When he threatened to kill them all, they bared their necks, and Pilate removed the standards.[18] Later, on the threshold of violent revolution, the high priests and Pharisees urged peace, warned those advocating armed revolt of the consequences, and defended the tradition of welcoming foreigners and their offerings in the Temple.[19] N. T. Wright argues that this was, however, a minority position among the Pharisees prior to the Great Revolt of 66–70 CE.[20]

Shortly after Jesus returns to Galilee from his ministry among the Gentiles, he leads his disciples to the villages near the city of Caesarea Philippi (Mark 8:27).

He reviews his work of healing and reconciliation that leads Peter to declare that Jesus is the Messiah. In response, Jesus prophesies that he will be rejected by those in power and put to death (8:31). From that point the narrative is punctuated with a series of locations that bring him closer and closer to Jerusalem. As he makes his way there, he teaches his disciples about the proper role of money, sex, and power in a world that conforms to God's law. He intersperses his teaching with two more prophecies that include rejection, suffering, and death (9:31 and 10:33–34).

Jesus's journey to Jerusalem is a deliberate move to assert a claim to the city and its sanctuary that will result in his death and apparent defeat. Jesus's execution will appear to his disciples and to the wider world as the defeat of his mission. Peter objects that as Messiah, Jesus cannot die (8:32–33). By refusing to defend himself with violence, he does risk the failure of his mission. But if he becomes violent he will undermine his mission of breaking the hold of this final challenge to the rule of God. Jesus shares the conviction of the Maccabean martyrs (2 Maccabees 7:13–23) that if he is faithful to the will of God, he will rise from the dead. His resurrection will release upon his disciples a life endowed with power not to be intimidated by the threat of death. The re-establishing of the divine rule will take a quantum leap. Death will have no power to provoke people to violence or to speak the final word on those who are faithful to the rule of God (Mark 13:9–11).

Jesus's rejection of violence was a rejection of the way of violent revolution pursued by the "thieves" who made the Temple their den several decades later. Their revolution failed. The Temple was destroyed as Jesus prophesied in his vision of the future he shared with his disciples while they gazed at this wonder of the world from the Mount of Olives.[21] Jesus's revolution did not fail.

## CHRISTIANITY AND VIOLENCE

For Israelites, resurrection was not an individual event. It was a component of a larger scenario that inaugurated a new heaven and new earth. To believe that Jesus had been raised from the dead meant that the work of lifting the curse of violence that Jesus had begun in his ministry had reached a climax: God had overcome death itself. From his followers' experience of Jesus's resurrected presence, the gifts of healing he brought the world during his earthly ministry continued to flow.

The rise of Christianity brought revolutionary change to Roman society. This included new attitudes toward the individual person, work, family, community, and the state itself. It transformed Roman society's view of God, introduced a new understanding of physical sickness and healing, a new use of money in acts of generosity, and new respect for private property.[22] In the first three centuries, converts to Christianity also followed Jesus's way of rejecting violence as a way of defending themselves against persecution. The stories of the early Christian martyrs portray them as following Jesus's teaching and example. There are no stories from this period that speak of Christians responding to acts of aggression with violence or retaliation.[23]

But near the end of the second century a new factor enters the picture: Christians in the military. The early Christian theologian Origen (184–253 CE) cites his pagan opponent Celsus's observation that prior to the end of the second century Christians by and large avoided military service.[24] But archaeologists have turned up tomb inscriptions for Christian soldiers for the period prior to Constantine. And beginning in the late second century, several prominent Christian writers object to Christians serving in the military, strongly suggesting that this was beginning to be an issue.

Initially these Christians were probably soldiers who had converted while serving in the army. Clement of Alexandria, who died in 215 CE, gives explicit approval to soldiers who remained in the military after their conversion. Tertullian (155–240 CE), Origen, and Hippolytus of Rome (180–230 CE) all object to Christians participating in bloodshed. Tertullian asks how a person can have an occupation involving violence when Jesus declares that the one who uses the sword shall die by the sword. Origen is clear that Christians cannot engage in warfare even for a just cause. Hippolytus declares that a soldier who wants to become a Christian must resign from the military. No Christian writer before Constantine explicitly defends military service.

Christians in the first three centuries were not small, persecuted communities trying to keep a low profile. They were an increasing presence in the population. They had been participants in society before conversion, often in positions of responsibility. They were susceptible to new social developments. Society was increasingly militarized and joining the army may have provided a path to upward mobility. As converts, some might have been instructed to withdraw from any position that would require them to kill or be otherwise violent, as they were by Hippolytus. But with Israel's scriptures as part of the Church's canon, they could have found support for a decision to remain in military service. By the end of the second century, it is plausible that a new, hybrid thinking emerged that made military service a legitimate occupation for Christians. But the legitimation of violence by Christian soldiers even after Constantine did not lack critics, especially among the monks. Sulpicius Severus (360–425 CE) and Paulinus of Nola (354–431 CE) speak against the participation of Christians in the military.[25]

## UNPRECEDENTED RESPONSIBILITY

In the fourth century, the Emperor Constantine embraced the Christian faith of his Greek mother, Helena, and ordered an end to government persecution of Christians, which had reached a high point in the reign of Diocletian, Constantine's immediate predecessor. Constantine also entered into a partnership with the so-called orthodox[26] Church to help bring unity and peace to the empire. This meant that the violence of the state was no longer directed at Christians. Quite the opposite: Christianity was now allied with a government that maintained order and protected itself from external enemies by violent means.

Now that it was in a position to influence the state, how should the Church, committed to Jesus's clear teachings on nonviolence, respond to government's use of military force? Up to this point, Christians could only respond to violence directed against them.[27] They lacked the means to defend other targets of the state. But under Constantine a new situation arose in which they shared the power of the state. Would they not be obliged to direct the state in shaping a Christian social order and to defend that order against violent attack?

American Catholic theologian William Cavanaugh notes that Jesus never pictured his Church in the situation it faced under Constantine, in which it both had an obligation to defend the powerless and possessed the violent means to do so. Cavanaugh suggests that the turn away from pacifism toward an accommodation of violence was far more complex than a simple move from fidelity to apostasy. New situations present new challenges and the situation called for creative response. The early Church, he writes, embarked on "the long pedagogy of God's people that did not end with the Advent of Jesus Christ, even though Christ showed us the definitive shape of the story's end."[28]

In his 2014 book, *Mercy: The Essence of the Gospels and the Key to Christian Life*, German liberal Roman Catholic theologian Cardinal Walter Kasper writes that "as paradoxical as it sounds, defense of one's own fundamental human rights or those of innocent individuals, especially women and children, against aggressive force and oppression, can be an act of neighborly love and, under certain conditions, an obligation of love of neighbor, provided that all other means have been exhausted and proportionality is maintained."[29] The context of Kasper's compromising of Jesus's teachings of nonviolence reflects the changed position of the Church in relation to governmental power. The principle of nonviolence, in his view, must be balanced against the obligation to defend the weak against the powerful.

But if the categorical rejection of violence is not appropriate to all situations in which Christians find themselves, neither is the categorical rejection of nonviolence as simply "unrealistic." The problem is that down through history many Christians have assumed that violence is the first, indeed only, possible response to conflict. There is little recognition of a tension between the sometimes unavoidable necessity of defending the vulnerable by violent means and Jesus's commitment to nonviolent resistance in both his teaching and his action. Jesus's command that his followers should take up their own crosses in imitation of him is not included in this understanding of the Christian religion. Jesus's commitment to nonviolence has been excised from Israel's story of God's commitment to healing creation and inserted into a different story, the Platonic narrative of individual, disembodied souls escaping the material creation for eternal life in a purely spiritual realm. Within this new story Jesus's self-sacrifice was reinterpreted as an act unique to him. People came to understand it as a source of forgiveness required if they were to spend eternity in heaven. It ceased to be a model for Jesus's followers to imitate as a realistic strategy for healing a fearful and divided world.[30]

With the key ingredient of Jesus's mission for healing the world transfigured to serve a story about abandoning this world for another, it has been easy for nations that claim Christian values to rationalize the use of violence as a way to cleanse the world of evil. With no significant countervailing commitment to find creative alternatives to aggression, Christianity's core teaching has been swallowed up by the seemingly limitless human capacity for violence. Authoritarian leaders and their followers might pay lip service to Christian values, but they play a zero-sum game. The enemy has no humanity, no legitimate case, no capacity for dialogue and compromise. To embrace the enemy, as Jesus did, is unthinkable. Obliteration, complete and total domination, is the only course of action. Without the modeling of nonviolence as a countervailing force in this equation, Jesus might just as well have never lived.

We need a radically different set of rules for how to treat each other and the Earth from which we draw our life. In Mark's story, Jesus's healing redemption does not occur in some far-off ideal time, in a vague future state of utopia. Mark's entire narrative is the story of the final "eschatological" transformation. But the narrative is also the story of how that eschatological transformation is no longer a far-off vision. In the life of Jesus, in the evolving ministry of his followers, and in all people of goodwill it has now begun to break in upon the world. Taken together, these forces have the potential to engender fields of compassion that assume palpable form in cumulative acts of healing, mercy, and reconciliation.

Jesus's way is emphatically realistic. It is the world's default strategy of more violence in response to violence that is unrealistic—and indeed toxic. Unchecked aggression as a first response has proven down through history to be disastrous.

In our own time, violence and cruelty have metastasized to levels the Genesis narrator could not have imagined. War has become asymmetrical, global, and constant. Acts of terror and political extremism have become facts of life. Violence drives business plans and political campaigns. Lethal weapons once reserved for combat have become tools for the paranoid, angry, or delusional to dispatch perceived enemies, inflicting harm and heartbreak. We have weaponized nature itself, mining fossil fuels from the Earth to run our vehicles and warm the planet in the process. Christianity itself has played a role in this destructive impulse. The false promises of the Prosperity Gospel rationalize greed and selfishness. The doomsday scenarios of dispensational theology encourage Christians to believe that the Earth is theirs to spoil, as they await an apocalypse that will send a faithful few to heaven. Authoritarian leaders outshout the apostles of peace; political operators manipulate our news and social networks with divisive messages and false reports.

CHAPTER 5

## TRANSFORMING LIFE ON EARTH

Cosmological theology directs us back to the matter of how to transform life *in this world* rather than in some distant and disembodied sphere of reality called "heaven." Jesus's work was this-worldly; it sought to redirect humanity from the wilderness of self-destruction back onto the path toward life. We long for a reversal of the radical evil that grips the human species with dire consequence for our entire planet. Theologians like Teilhard de Chardin, who have brought Christian theology into serious dialogue with modern science, even envision that the universe really is evolving to the incarnate body of Christ. We as a species equipped with a primitive fear have developed technology so powerful that by using it we will destroy ourselves. Many people of all religions and no religion are doing what they can to make life better where they can. They resist the temptation to be possessed by fear and strive to love. Will the free and creative human spirit evolve toward love in time to avert catastrophe?

Stephen Jay Gould, noted paleontologist, evolutionary biologist, and historian of science, was an agnostic, but his view of the transformative value of compassion opens a window to the way the teachings of Jesus can have relevance even to those who are not among his followers. Gould wrote this in response to the 2001 terror attacks on the World Trade Center:

> The tragedy of human history lies in the enormous potential for destruction in rare acts of evil, not in the high frequency of evil people. Complex systems can only be built step by step, whereas destruction requires but an instant. Thus, in what I like to call the Great Asymmetry, every spectacular incident of evil will be balanced by 10,000 acts of kindness, too often unnoted and invisible as the "ordinary" efforts of a vast majority.[31]

The Genesis narrator traces the origin of violence rising from the murderous impulses of Cain and Lamech's poisonous acts of vengeance. This impulse to aggression has been incorporated over time into humankind's behavioral DNA. In *Freedom, Suffering and Love*, British clergyman Andrew Elphinstone observes that these impulses have been there from the beginning of human evolution. In light of what modern science has been discovering about human origins, Elphinstone concludes that there never was a literal Paradise. From the beginning, we possessed behaviors designed to protect ourselves in response to our primitive fear of "the other." The Genesis myths do not accurately explain how we became violent, but they reveal insights about our violence.[32] In the shadow of the abandoned tower of Babel, Noah's power-hungry descendants were made, because of their arrogance, strangers to one another. No longer bound by a common language, a shared sense of human identity shattered, they push farther into the wilderness. They fear strangers rather than welcome them, striking out in mutual expressions of distrust and aggression.

According to Elphinstone, Jesus introduced into the story of human evolution a new option: addressing fear and violence with love. Love is a latecomer in human evolution. It challenges all our evolved instincts. Elphinstone interprets Israel's narrative of the cosmos as a history of learning the lesson of love. Love makes us vulnerable, and when we feel vulnerable we instinctively defend ourselves, most readily with violent words and violent actions. Yet when we strike out against those we fear instead of learning to live with them in peace, we perpetuate the cycle of violence. In the past our acts of violence served us well as we struggled against all that threatened us. But we have now developed means for defending ourselves that are so powerful that if we employ them we will destroy ourselves along with our enemies. Love is now our only hope if life is to continue on Earth.

The forces that threaten our self-destruction are formidable. These patterns traced in Genesis's ancient myth of origins continue today as nations slam their borders shut to refugees fleeing war, as terrorists wreak vengeance against the innocent, and as autocratic leaders recklessly exploit their power. But there is hope. Modern science yields a picture of the universe as one great, interconnected system in which even very small events can have world-changing consequences. Meteorologist Edward Lorenz tells us small actions can result in great changes: the flapping of a butterfly wing on one side of the planet can affect the weather pattern on the other side of the Earth. Jesus began with himself and then called a small group of disciples—women as well as men, Mark surprisingly informs us—to a life of love in place of violence. The little he and they were able to accomplish caught on with an ever-widening circle of people and is still a force today. What Jesus and those faithful to his way are practicing is not limited to them. We witness nonviolent acts of resistance motivated by love among people of all faiths or of no religious faith, as in the case of Stephen Jay

Gould. The call to love, a new chapter in the evolution of the human species, is a powerful message with universal appeal. All those who embrace this idea of a collective compassionate impulse—singular acts of mercy that seek to heal the wounded and embrace the stranger—can work together to profoundly transform a broken world.

# 6

# Visionaries, Prophets, and Saints

In 2013, not long after he was elected to lead the Roman Catholic Church, Pope Francis described an arresting vision of how people of goodwill might respond to a world cursed by violence, riven by suffering, and threatened by environmental ruin: the field hospital—a metaphorical tent erected in a crisis zone, staffed by empathetic individuals poised to embrace the suffering stranger.

> [T]he thing the church needs most today is the ability to heal wounds and to warm the hearts of the faithful; it needs nearness, proximity. I see the church as a field hospital after battle. It is useless to ask a seriously injured person if he has high cholesterol and about the level of his blood sugars! You have to heal his wounds. You have to heal the wounds. Then we can talk about everything else. Heal the wounds, heal the wounds. . . . And you have to start from the ground up.[1]

The field hospital metaphor summons images of altruists who instinctively rush toward danger to ease suffering and save human life: The pacifist stretcher-bearers in the trenches of World War I. Doctors Without Borders fighting famine and disease around the world. First responders who rush to the scene of a tragedy. Journalists who bear witness to brutality and injustice. Activists who protest injustice and stand up for the vulnerable. Bodhisattvas who vow to ease suffering for every sentient being.

Sexual misconduct and abuse of power occur in virtually any religious hierarchy, and have turned many away from organized religion. Increasing numbers of people describe themselves as spiritual, not religious. The field hospital metaphor is an appealing antidote: an inclusive and altruistic community that is a process rather than an institution: a place of healing and reconciliation, powered by individual acts of compassion.

At a time when evil is often expressed in the faceless actions of authoritarian governments and institutions, to serve in the field hospital makes a real human difference. To respond to disaster with individual acts of healing—person by person, wound by wound. These islands of calm in the chaos have cosmic significance. To stand in solidarity with those who suffer is to speak the truth and do the needful in daunting circumstances. In the bleak moral urgency of the present, this kind of courageous compassion can be a useful and necessary instrument to guide a wounded world away from evil and back toward the original harmony that the creation myths of Genesis so vividly describe.

What does it take to serve in the field hospital of contemporary life? How do ordinary people cultivate courageous compassion for the benefit of all? How can human beings evolve, individually and collectively, particles in an ever-expanding wave of compassionate energy that binds us together in peace?

Every era has its visionaries, prophets, and saints: people who take heroic action in response to urgent moral crises. Inspiration from role models can be a first step. But "role model" is a tricky concept. Saints frequently sin. Visionaries occasionally need corrective lenses. The words of prophets are often ignored.

The following portraits examine pivotal moments in the lives of remarkable people who brought a singular courage, grace, and insight to the times in which they lived: the cosmological theologian Pierre Teilhard de Chardin, the prophetic partners Martin Luther King Jr. and Abraham Joshua Heschel, and Dorothy Day, peace activist and servant of the poor. Rabbi Heschel's daughter, religion scholar Susannah Heschel, also figures in his story. Each individual was a fully human being, vulnerable to fate and moral failure. Each faced challenges that were as daunting as the moral and environmental crises we face today.

## PIERRE TEILHARD DE CHARDIN: A MYSTICAL VISION

Pierre Teilhard de Chardin was a scientist and a mystic. Like Pope Francis, he was a Jesuit priest. His writings reflect a belief that the divine is discoverable in the majesty of the universe. He saw the evolutionary interplay of matter and spirit as a force that propels human consciousness forward into the future.

In his view, the material world is not debased, it is sanctified. The religious experience is not merely personal, it is cosmic. Genesis is a process and the universe is a liturgy, imbued with eucharistic holiness. God's transformative presence is discoverable in the energies of the cosmos, expanding into the future, pointing to an inevitable union with the divine.

His vision of evolution was both spiritual and material, fueled by an ever-expanding human consciousness that binds us together in a disembodied network of shared knowledge. This network, which he called the noosphere, foretold the best of what can happen in the disembodied realms of cyberspace, where individual minds can expand, meet, merge, and diversify in an evolutionary dance.

In the early twentieth century, these beliefs were considered heretical, but today Teilhard de Chardin's work is embraced by the Catholic Church and many other faith groups seeking a new way of thinking about spirituality, science, and a meaningful response to the global environmental crisis.

Teilhard de Chardin was born in 1881 in France to a pious and prominent family; his aristocratic father was an amateur naturalist; his mother, a niece of Voltaire, influenced him with her Christian mysticism and profound love of the natural world. A childhood fascination with the wonders of nature led him to the priesthood and to the study of geology, zoology, and paleontology. His spiritual advisors encouraged him to become a scientist.

When World War I broke out, the young seminarian left his scientific studies and joined the French army as a noncombatant, serving as stretcher-bearer for a combat battalion of North African Zouaves. The savagery of the conflict that killed 20 million and wounded 17 million more tested his faith. Yet out of the depths of this devastating global conflict, he was able to discern a more complex pattern and a higher purpose: He believed humankind had the potential to evolve itself away from the evil. He could see the universe expanding in the embrace of a loving God, pointing toward an "Omega point" of unification of the material and the divine.

In 1916, during the battle at Verdun, he described it this way in a letter to his cousin Marguerite:

> I don't know what sort of monument the country will later put up on Froideterre hill to commemorate the great battle. There's only one that would be appropriate: a great figure of Christ. Only the image of the crucified can sum up, express and relieve all the horror, and beauty, all the hope and deep mystery in such an avalanche of conflict and sorrows. As I looked at this scene of bitter toil, I felt completely overcome by the thought that I had the honour of standing at one of the two or three spots on which, at this very moment, the whole life of the universe surges and ebbs in places of pain but it is there that a great future (this I believe more and more) is taking shape.[2]

For multiple acts of heroism on the battlefield, the government of France awarded Teilhard de Chardin the Croix de Guerre. In the trenches, as he tended to the wounded and dying, he considered what the shell-shocked world might become, if humanity had the capacity to evolve beyond the battlefield to a higher plane.

After the war, Teilhard de Chardin continued his studies at the Institut Catholique in Paris. He was influenced by the French philosopher Henri Bergson, who rejected the prevailing dualistic view of a world composed of matter and spirit in favor of a continuously expanding "tide of life." Bergson saw this life force as undirected by any higher power, but Teilhard de Chardin eventually developed his own spiritual and transformational interpretation of the origins and direction of the universe.[3]

His theological writings were dismissed by church authorities who viewed evolution as a godless, dangerous, and materialistic view of the world. Bergson's *Creative Evolution* was placed on the Vatican Index of Forbidden Books. The Vatican's theological gatekeepers attacked Teilhard de Chardin and others who suggested a spiritual dimension to evolution. By 1925, conservative French bishops called for a Vatican investigation, alarmed by the charismatic young scholar whose mystical reading of evolution was at odds with church dogma on original sin.

Teilhard de Chardin was told to repudiate his ideas about evolution. He then embarked on a life of scientific discovery, joining an international team of scholars exploring archaeological sites in China. He participated in the expedition that resulted in the discovery and interpretation of the prehistoric fossils and artifacts of the 750,000-year-old Peking Man. His explorations continued in India and Malaya in the 1930s and 1940s and later as a member of major archaeological expeditions in the Middle East, South America, and Africa.

The Vatican's ban on publishing or speaking in public on theological matters caused him much anguish, yet Teilhard de Chardin observed a public silence and remained a faithful priest. But he did not refrain from criticizing the institutional church, which he considered to be prejudiced against women, stuck in the past, ignorant of science, and unable to respond to the realities of life on Earth. "It has sometimes seemed to me there are three weak stones sitting dangerously in the foundations of the modern Church," he wrote to a friend. "First, a government that excludes democracy; second, a priesthood that excludes and minimises women; third, a revelation that excludes, for the future, prophecy."[4]

In 1929, he encountered the American sculptor Lucile Swan in China, marking the start of an intimate friendship that continued for twenty-five years. Swan's published letters provide a view into a creative relationship between a priest and a woman that is uncommon in the Catholic priesthood.[5]

Beyond the Vatican, Teilhard de Chardin had many admirers and plenty of critics. Secular scientists dismissed his mystical reading of evolution. Yet many were drawn to this enigmatic figure who concealed behind a veil of personal piety a great passion for life, a sense of mystery, and a sense of humor. "He was a passionate man," the paleontologist Stephen Jay Gould wrote in 1983, "a genuine hero in war, a true adventurer in the field, a man who loved life and people, who strove to experience the world in all its pleasures and pains."[6]

When Teilhard de Chardin died in New York City in 1955, the Vatican ban on his religious scholarship became ineffectual. His theological works were soon disseminated by secular publishing houses and were translated into many languages. *The Phenomenon of Man* became an international best-seller. And for the first time, his work came under greater scrutiny in the scholarly world.

From the perspective of contemporary theological studies, which value diversity and an ecumenical spirit, his Christocentric view of evolution may seem insular.[7] His enthusiasm for evolution as the engine of perfection raises ethical questions about natural selection itself. What about evolution's losers: the weak, the impaired, the vulnerable?[8]

Teilhard de Chardin was not an environmentalist. It has been said that he had more appreciation for the design of an airplane than for the miracle of a bird. He was so captivated by the idea of progress that he did not readily see its negative impact on the natural world. He was thrilled by the idea of nuclear power and his advocacy of unrestrained experimentation in science and technology carried alarming overtones of the eugenics policies that would stain the history of the twentieth century. He died before the widespread damage human beings have done to the environment became unavoidably evident.

That his ideas are now identified with the environmental movement is largely due to Thomas Berry, who was a major interpreter of Teilhard de Chardin's work. Over a lifetime of scholarship and commentary, Berry acknowledged the French paleontologist's ecological blind spots and expanded his ideas to include environmental concerns. In Berry's view, Teilhard de Chardin's impact on Christianity was as great as St. Paul's. By telling a story of evolution that was at once material, psychic, and spiritual, one in which human concerns were embedded, Teilhard de Chardin offered a new way of experiencing the divine. He moved the essential Christian issue from salvation to creation.[9]

Many conservative Catholics still consider Teilhard de Chardin's work heretical, but the official church that originally rejected him has now embraced his writings. John Paul II, Benedict XVI, and Francis all have cited his vision of the sacred and evolving cosmos as a guide for church teaching on the environment. Episcopalians in the United States and Anglicans worldwide observe a feast day in their liturgical calendars.

At a time of political chaos, environmental upheaval, and digitally driven social dysfunction, Teilhard de Chardin's ideas about spiritual and psychic energy, science, and technology have fresh meaning. His ideas have value not only to people of faith, but to the growing cohort of those who consider themselves spiritual but not religious, who find transcendent meaning in the beauty of science and the awesome mysteries of the universe.

His vision took shape in the darkest of hours. Under devastating circumstances in a savage war, a noncombatant stretcher-bearer was able to take the long view of evolution as an inexorable spiritual and material process in which humans are full participants, a force so relentless that it has the potential to evolve beyond evil itself. And his concept of the "noosphere" could serve as a prototype for evolving our way out of the dysfunctional cloud of information and social media in which we dwell today. Teilhard de Chardin envisioned something quite different from what Facebook, Apple, and Google have given us: a disembodied, ever-expanding cloud of knowledge that binds humankind together and propels us. United in compassion and empowered by our shared knowledge, humanity adapts to circumstances as they arise and moves forward into the future, in an ever-expanding universe, imbued with spiritual and psychic energy and the force of the divine.

## ABRAHAM JOSHUA HESCHEL AND MARTIN LUTHER KING JR.: PROPHETS ON A BRIDGE

There is a tradition in Hasidic Judaism, rooted in the teaching of the prophets, that God needs humans as much as humans need God. On all of our dangerous journeys, God walks beside us; God feels our pain; God shares our sorrows and our joys. And when we do harm to others, we also do harm to God.

Jesus was born a Jew and we know from the study of Mark's gospel that his first followers were Israelites who retained their religious and tribal identity as they sought to follow his teachings. But the tapestries of Christian history are threaded with anti-Semitism; the libel that Jews were responsible for the killing of Jesus has rationalized pogroms and violent persecutions over centuries. Condemnation of Jews and hope for their ultimate conversion have been knitted into Christian readings of the scriptures and prayer practices.

The rise of Adolf Hitler left a stain on Christianity as it was practiced by ordinary Germans of the time. As the Third Reich accumulated power and began a systematic persecution of Jews in all levels of society, a group of German Protestant theologians, bent on solidifying their status with the regime, inflicted significant harm on the idea of God: They undertook a project to expunge the Hebrew scriptures from the Bible and recast the historical Jesus into an artifact of Aryan descent, distorting his teachings of nonviolence, reconciliation, and healing to conform with Nazi ideology.

These scholars declared that Jesus was born an Aryan. In language that prefigured the rhetoric of twenty-first-century racists and anti-Semites, the works of these Nazi-influenced academics cast Jesus as a reformer who would replace Judaism with a new and better religion. They argued that Judeans who pressured Roman authorities to crucify Jesus were combatants in a race war against Aryans; they claimed that the Christian spirit found its ultimate expression in the Aryan race. They taught that the Hebrew scriptures, its stories, psalms, prophecies, and wisdom, no longer had a place in the Bible and must be excised from the official text.

# CHAPTER 6

Though some contemporary scholars were aware of what occurred in some German schools of theology under the Third Reich, the average Christian was generally ignorant until 2008, when Susannah Heschel opened a window on that dark time, with her book, *The Aryan Jesus: Christian Theologians and the Bible in Nazi Germany*. Heschel was the first scholar to gain access to the archives of the Institute for the Study and Eradication of Jewish Influence on German Religious Life, an academic research center at the University of Jena. Its director was New Testament scholar Walter Grundmann.

Grundmann's institute published a version of the New Testament with all references to Judaism expunged. The institute convened conferences and published books that defamed Judaism, receiving support from bishops, clergy and academics. This filtered down into Sunday sermons uttered from the pulpits in German churches, abetting the spread of poisonous racist and anti-Semitic ideology into ordinary people's lives. While there is no evidence that directly links these faithless clerics and theologians to the heinous crimes of the Holocaust, Susannah Heschel makes clear the particular evil of distorting religious teachings and deploying them as a weapon:

> One cannot prove that the Institute's propaganda helped cause the Holocaust. However, the effort to dejudaize Christianity was also an attempt to erase moral objections to Nazi anti-Semitism. Institute-sponsored research, by describing Jesus's goal as the eradication of Judaism, effectively reframed Nazism as the fulfillment of Christianity. Whether the Nazi killers of Jews were motivated by Institute propaganda cannot be proven, but some did express gratitude for Institution publications, apparently for alleviating a troubled conscience. Institute publications were not as widely disseminated as the propaganda issued by the Reich Minister of Propaganda, Joseph Goebbels, or the publications of Julius Streicher, who was hanged at Nuremberg for editing *Der Sturmer*, a weekly anti-Semitic propaganda rag. Yet the moral and societal location of clergy and theologians lends greater weight to the propaganda of the Institute; propaganda coming from the pulpit calls forth far deeper resonance than that spoken by a politician or journalist.[10]

In the war-crimes tribunals of the postwar years, the faithless theologians and clergy largely managed to escape official scrutiny. Some, like Grundmann, actually flourished; he achieved popularity in the 1950s as a conservative evangelical preacher. Seen against the magnitude of the Holocaust, Grundmann and his colleagues might appear to be minor players. But their moral offense is huge: In the hearts and minds of believers, they damaged the image of God.

Susannah Heschel had good reason to explore this topic. Her father, Abraham Joshua Heschel, is recognized as one of the most influential religious leaders of the twentieth century. He was a rabbinical student in Germany at about the same time that Grundmann and his colleagues were rewriting the New Testament. Heschel managed to leave Germany in 1939 and ultimately settled in America; the members of his family who remained perished in the Holocaust.

In the years after World War II, Heschel, a passionate and charismatic proponent of social justice, civil rights, and inter-religious reconciliation, was known as "the people's rabbi." Descended from a long line of socially conscious Hasidic religious leaders, he spoke and wrote eloquently about Jewish values and mysticism. He was a skilled interfaith diplomat who stood as a bridge between Christians and Jews, engaging in deep conversations with Pope John XXIII about reconciliation between Catholics and Jews. He played a significant role in the church's decision to remove anti-Semitic references from Catholic scripture and liturgy.[11] A prophetic consciousness of the deep and intimate relationships between humans and their creator animated his life and work:

> To the prophet . . . God does not reveal himself in an abstract aloofness, but in personal and intimate relation with the world. He does not simply command and expect obedience; He is also moved and affected by what happens in the world and reacts accordingly. Events and human actions arouse in Him joy or sorrow, pleasure or wrath. He is not conceived as judging the world in detachment. He reacts in an intimate and subjective manner, and thus determines the value of events. Quite obviously, in the biblical view, man's deeds may move Him, affect Him, grieve Him or, on the other hand, gladden and please Him. This notion that God can be intimately affected, that he possesses not merely intelligence and will, but also pathos, basically defines the prophetic consciousness of God.[12]

Susannah Heschel reflected on her father's understanding that we are not merely bystanders in the cosmos, but part of the unfolding dynamic that binds people of faith to each other and to God:

> It goes back to rabbinic Judaism and to the prophets. There is a divine need. My father calls it "divine pathos." That need is for us to observe the commandments,

> for example, to help God achieve redemption. It's not only we who are in exile, but God is in exile. Classically, in Jewish thought, we help God achieve redemption through observance of the *mitzvot*, of the commandments, keeping the Sabbath and prayer.
>
> My father expanded that understanding to include also commandments that involve care for other people. His concern was with all human beings. "Never be indifferent to other people's suffering" was the way he concluded a brief talk about the Holocaust. He didn't say, "Make sure nothing like this happens to *us* again"—make sure it never happens to *anybody* else, ever.[13]

When Abraham Joshua Heschel first encountered Martin Luther King Jr. in 1963, it was a case of one prophet recognizing another. King, a Baptist minister, was born in the Jim Crow South, schooled in Protestant theology and Gandhi's ideas of nonviolence. Heschel, born in Warsaw, was a disciple of Martin Buber, held a doctorate in the study of the prophets from the University of Berlin, and was a moving force in liberal Judaism. Both spoke with the passion of prophets, challenging the evils of racism, economic injustice, and the immorality of war. Both rejected the idea, rooted in Greek philosophy, that God is a distant presence, unaffected by human action. Both embraced the idea of a God pained by our cruelties, accompanying us in our exiles, engaged in our quest for freedom. Heschel came from a world in which Protestant theologians tried to erase Judaism from the Bible. King lived in a world that tried to erase the dignity of African Americans and their demands for equal treatment under the law. King was an Old Testament preacher; Exodus was the metaphor of the Civil Rights movement; his lodestars were Moses, Amos, and Isaiah. Both King and Heschel were accustomed to speaking truth to power.

At the time King first drew Heschel into the ranks of religious leaders in the forefront of the Civil Rights movement, the American Jewish community was as divided as the rest of the country. In the Freedom Summer of 1964, the murder of three civil-rights workers in Mississippi was a galvanizing moment: James Chaney, an African-American voting-rights worker, and two Jewish volunteers from New York, Andrew Goodman and Michael Schwerner, were murdered on their way to investigate the burning of an African-American church.

Less than a year later, on March 7, 1965, a peaceful civil rights march turned into a bloody confrontation with Alabama state troopers who wielded whips, nightsticks, and tear gas. A second march, two days later, was short-lived. The

Reverend James Reeb, a Unitarian minister who had joined the demonstration, was beaten to death by white assailants, sparking a national outcry and calls for massive civil disobedience.

King resolved that the marchers would not be deterred. He sent out an urgent call to Heschel and other religious leaders to join him on a third march from the bridge all the way to the state capitol in Montgomery, more than fifty miles away.

Susannah Heschel, then only a child, absorbed the carnage of the first confrontation on the Edmund Pettus Bridge as most of America did, on a tiny black-and-white television screen in her family's Manhattan apartment. The flickering images of police dogs lunging at demonstrators summoned comparisons with what happened to the Jews of Germany a generation earlier. And when the telegram came from King asking Heschel to stand with him in Selma, Susannah was fearful. Her father was not in good health; the seminary where he was teaching did not approve of his decision to cancel classes to attend the march. She wondered if she would ever see her father again.[14]

When Rabbi Heschel arrived in Selma, he stayed at the home of Dr. Sullivan Jackson, an African-American dentist who often hosted King and his followers. On the morning of the march, they assembled in the living room, each preparing in his own way for whatever lay ahead. King was in one corner of the living room, praying. Heschel was in another corner, praying. A Catholic priest was elsewhere in the house, also praying. The mood was tense; they did not know what violent confrontations might await them. But it was also somehow festive. Many of the marchers, King and Heschel included, wore flower leis that had been sent as a blessing from supporters in Hawaii.

King, the Old Testament preacher, set forth from the Jackson house to the Edmund Pettus Bridge with Heschel, the learned rabbi and refugee of war, to join some two thousand other marchers, this time protected by federal troops dispatched by President Lyndon Johnson. Dazzling in their diversity, the marchers included Catholics, Protestants, Orthodox Christians, students, activists, diplomats, political leaders, people of great religious faith, and people of no faith at all. They moved as one organism, particles in a great wave, insisting on human dignity, insisting on a political change, insisting that America evolve beyond its history of racism and divisiveness into a more just and equitable society.

Later, in his diary, Heschel would recall their three-day journey to Montgomery under watch of federal troops, sleeping in fields, drawing the nation's attention to their protest. "Legs are not lips and walking is not kneeling," Heschel wrote. "And yet our legs uttered songs. Even without words, our march was worship. I felt my legs were praying."[15]

By the time the original two thousand demonstrators arrived at their destination at the state capitol, the crowd had swelled to twenty-five thousand, demanding legislation that would guarantee all Americans the right to vote. And because of their efforts, America would cross a bridge to a new era. The Voting Rights Act that resulted from their actions would not resolve the legacy of bigotry and divisiveness that continues to this day. But prodded by these two charismatic prophets, Heschel and King, America experienced a holy moment in its troubled history. People of goodwill had joined together, bridged their divisions, and met violence with nonviolence. That moment did not last. But it stands as an inflection point in which prophets of two great religious traditions showed that it was possible to repair the damage that human hatred had inflicted on the world and its Creator.

## DOROTHY DAY: EMBRACE AND RESIST

There was always a certain aura around Dorothy Day, a force field that emanated from her long public life as a pacifist, political activist, advocate for social justice, and servant of the poor. It was more than her arresting physical beauty or her personal charisma as leader of the progressive Catholic Worker movement, or her implacable opposition to all forms of war. It was more than the hard-edged wisdom of her autobiography, her writings on spirituality, or her condemnation of the morally bankrupt ideologies of her times. It was more than the vigor with which she embraced a life of intentional poverty, more than her talent to see beauty both in nature and in the midst of squalor and distress.

There was a holiness about her, a spiritual force nurtured by a commitment to social justice. Her religious writings and private meditations, letters and diaries, personal prayers, and public witness all resonate with a love for the natural world, an unrestrained sense of joy, and the conviction that individual works of mercy can heal a wounded world.

These are attributes of those we call saints. It is ironic that the Catholic Church has cranked up the antiquated bureaucracy that would declare Dorothy

Day a saint. A seal of approval from the institutional church was a notion this perpetual outsider scoffed at in her lifetime, not only because she was a sharp critic of the clerical hierarchy. She believed that we are all called to be saints.

Day was, at various points in her life, a bohemian, a socialist, an anarchist, an atheist, an activist, a communist, a radical, a rebel. She lived in a state of permanent protest and willingly went to prison for her beliefs. She resisted paying taxes that would fund any kind of conflict. In addition to the homeless, she also sheltered draft dodgers and military deserters. The Federal Bureau of Investigation maintained a thick file of her activities. She had a short temper, a sharp tongue, and a tendency to think that the life decisions that worked for her would also work for others. She suffered from migraines and bouts of depression. For much of her life, she smoked incessantly.

Her first arrest, at twenty, in 1917, occurred outside the gates of the White House, while she was protesting the mistreatment of women jailed for demanding the right to vote. On a hunger strike in solitary confinement in a Washington, DC, jail, she experienced some of the same excruciating circumstances she had been protesting. Soon after, she dropped out of college to nurse victims of the Spanish influenza epidemic.

In New York in the 1920s, she worked as a reporter for newspapers like *The Call* (socialist) and *The Masses* (communist), wrote a potboiler novel about her bohemian lifestyle, and briefly ventured to Hollywood to write movie scripts. She ran with a hard-drinking crowd of anarchists and revolutionaries, writers, actors, and playwrights, including journalist John Reed, Katherine Anne Porter, and Upton Sinclair. Eugene O'Neill was an intimate friend; Day is believed to have inspired the sharp-tongued character of Josie in O'Neill's final play, *A Moon for the Misbegotten*.[16]

Day was a woman of passion, a reluctant celibate, haunted by God, thirsty for justice, hungry for love and for the generative stability of motherhood. Two love affairs resulted in pregnancies; one ended in abortion. Forster Batterham, the father of Day's only child, Tamar, was an anarchist and atheist who didn't believe in marriage. He supported Day's pregnancy, but was repulsed by her burgeoning interest in religion. Their union fell apart when she embraced Roman Catholicism. An unapologetic single mother at a time when this carried a great stigma, she spent years trying to convince Batterham that it was their shared destiny to marry and raise a large family together. This was an enduring heartbreak, but their relationship as parents and grandparents would continue for the rest of their lives.

The economic disaster of the Great Depression crystallized her calling. With the French agrarian philosopher Peter Maurin, she cofounded the utopian Catholic Worker movement. Combining progressive Catholic social teaching with a deep sense of spirituality, a belief in the healing properties of nature, and a dedication to personal service to the poor, Catholic Workers fed and sheltered thousands through a network of thirty-three shelters and farms. They pursued lives of intentional poverty, dwelling as equals in the same harsh conditions experienced by the people they served. Day traversed the country to shore up support for the enterprise, and while she was praised for her heroic actions for the poor, she was also criticized for leaving the task of raising Tamar to her Catholic Worker colleagues and the nuns in a series of boarding schools.

Day was editor of the *Catholic Worker* newspaper for nearly a half-century. The *Worker*, which sold on the street for a penny and was distributed free at working-class Catholic churches nationwide, was an initial success. By 1938, its circulation reached 190,000. But Day's unwavering opposition to violence worked against the success of the enterprise. One month after the attack on Pearl Harbor, Day published passionate arguments against joining the conflict:

> We are at war, a declared war, with Japan, Germany and Italy. But still we can repeat Christ's words, each day, holding them close in our hearts, each month printing them in the paper. In times past, Europe has been a battlefield. But let us remember St. Francis, who spoke of peace and we will remind our readers of him, too, so they will not forget.
>
> In The Catholic Worker we will quote our Pope, our saints, our priests. We will go on printing the articles which remind us today that we are all "called to be saints," that we are other Christs, reminding us of the priesthood of the laity.
>
> We are still pacifists. Our manifesto is the Sermon on the Mount, which means that we will try to be peacemakers. Speaking for many of our conscientious objectors, we will not participate in armed warfare or in making munitions, or by buying government bonds to prosecute the war, or in urging others to these efforts.
>
> But neither will we be carping in our criticism. We love our country and we love our President. We have been the only country in the world where men of all nations have taken refuge from oppression. We recognize that while in the order

> of intention we have tried to stand for peace, for love of our brother, in the order of execution we have failed as Americans in living up to our principles.[17]

Day's uncompromising pacifism did not sit well with patriotic, blue-collar American Catholics; it also polarized the Catholic Worker community. Subscriptions to the newspaper plummeted. As the nation marched off to war, the *Catholic Worker* was relegated to the fringes.

Her sense of theology was uncomplicated, grounded in the Catholic notion of the mystical body: a community of saints, living souls united in empathy, making a tangible difference in the world. Her personal religious practices were typical of the devout midcentury Catholic: daily Mass, a devotion to Mary, recitation of the rosary, novenas and spiritual retreats, and the firm belief that real-world problems could be solved by prayer to and imitation of the lives of the saints. She served as a spiritual model to generations of Catholic religious women and laity. In the 1950s, Day's family watched her fall under the influence of a conservative spiritual advisor who drew her into a subculture of rigorous spiritual retreats, self-denial, fasting, and detachment from the beauties of the world she had once so joyfully embraced. It was a phase that would eventually pass. In an illuminating 2017 memoir, Kate Hennessy, one of Day's nine grandchildren, provides intimate details about the experience of having a perceived saint in the family. After bearing nine children and enduring an abusive marriage, Day's daughter, Tamar, ultimately left the Catholic Church, but remained faithful to the Catholic Worker movement and its ethos of serving the poor. Day's grandchildren were largely indifferent to her devout Catholicism. Regardless of their differences, Tamar and her brood remained passionately attached to the woman they called Granny.[18]

Hennessy explained the gap between the public perception of Dorothy Day, charismatic servant of the poor, and the paradoxical woman she knew and loved in a 2017 conversation with book reviewer Shannon Hill:

> Often people who didn't know Dorothy would oversimplify her and her life, irritating my mother to no end. Separating Dorothy from her "sins" or her shortcomings does her, and us all, a disservice. These are essential elements of her life that helped make her who she was. Her formidable strengths came out of her failures and weaknesses. And if you examine the lives of the saints, you will find many with equally messy lives and impossible personalities.[19]

As the Catholic Worker movement struggled to keep its utopian focus, maintain its commitment to intentional poverty, and operate its network of homeless shelters and farms, Day continued her work on the frontlines of the Civil Rights, anti-war, and labor movements. She challenged the prevailing notion in Cold War America that nuclear weapons were a rational response to political conflict. Alarmed that routine duck-and-cover exercises in schools and public places would normalize the idea of nuclear war, she refused to participate in Civil Defense drills. She refused to pay federal taxes that would support any form of aggression. As a result, she was a frequent resident of the Manhattan Women's House of Detention and other prisons around the country. Her prayerful public protests in Rome during the first sessions of the Second Vatican Council in the 1960s helped push the Catholic hierarchy to finally issue historic statements that opposed the development of nuclear weapons, championed human rights, and professed solidarity with the poor.

In the field hospital of compassionate response to the moral crises of her times, Dorothy Day showed up early and stayed late, blending the fire of a social reformer with the grace of a deeply prayerful person. A remnant of the movement she founded survives today, quietly operating in small shelters and farms around the country, a community of faith committed to serving migrants, the homeless, the lost, and the vulnerable.

Day lived in times as hazardous as our own: Individual greed and political corruption widened the gap between rich and poor. Authoritarian leaders demonized minorities and used propaganda to mislead and manipulate the populace. Proliferating nuclear weapons made mutually assured destruction a real and continued threat. Her great gift as an activist was that her protests were prayerful, not angry—a useful tactic for our own era of divisive public discourse. Her even greater gift was a mystic's vision that found beauty in the most unlikely places: In the waste-strewn waters off her beachside cottage on Staten Island, she looked beyond the garbage to celebrate the power and predictability of tides; in the waves of desperate, damaged, and lonely people seeking hospitality at Catholic Worker houses, she could see the face of God.

Her last arrest, at the age of seventy-five, occurred in a sun-drenched orchard near Lamont, California, in 1973. Day took a stand with Cesar Chavez and the United Farmworkers Movement to demand fair wages and working conditions for the migrant workers of the Central Valley. *Catholic Worker* photographer

Bob Fitch captured the iconic image of an immovable septuagenarian, seated on a campstool, defiant under the shelter of a broad-brimmed hat.[20] Two California state troopers loom over her, holstered guns on their hips. She would soon be arrested with the striking migrant workers and sentenced to thirty days on a Central Valley prison farm. She holds the troopers in a steely gaze; she is reciting verses from the Sermon on the Mount:

> Blessed are the poor in spirit: for theirs is the kingdom of heaven . . .
> Blessed are the meek: for they shall possess the land . . .
> Blessed are the peacemakers: for they shall be called children of God . . .
> Blessed are those who hunger and thirst after justice: for they shall have their fill.

# 7

# The Seed and the Soil

In a world full of danger, injustice, and suffering, the human species faces a crucial spiritual, ethical, and moral task: to consciously and continually make a difference in a broken world. Each individual act of compassion can be seen as one particle in the wave of evolutionary energy that shapes the emerging cosmos. Each personal decision to promote healing and reconciliation continues the work that Israel initiated after the fall of Adam and Eve and that Jesus continued in the Gospel of Mark: healing the rift between one another, the natural world, and the Creator. Each positive effort continues the process of restoring our lost partnership with God, as we participate in the awesome unfolding of the universe.

Jesus's first public teaching session, as reported in Mark's fourth chapter, uses an agricultural metaphor to describe his ministry on Earth and the challenges that await those who follow him. Lutheran theologian Virgil Thompson has observed that this one story could explain Jesus's entire mission.[1]

Jesus likens the different people who have heard his teachings to different kinds of dirt: a beaten-down path where seed never takes root; a shallow layer of soil where plants wither in the heat of the Palestinian sun; tillable soil whose seedlings are choked by thorny branches; and a few patches of deep, fertile ground that yield a crop beyond belief (Mark 4:1–9).

We might wonder what to make of this story. Mark lets us listen in as Jesus explains it to his disciples. If we don't get the meaning of the metaphors, we are in good company. They didn't get it either.

Are there hardened people among these listeners who are like a well-trodden path on which the seed just bounces when it falls? Are there any who are affected by Jesus's words and deeds and joyfully receive them, but whose response does not go very deep? Are there others whom the message penetrates, but who are always so controlled by their issues that they fail to respond? Are there any among these listeners who are like the deep, rich soil in which the Kingdom of God can flourish?

If all these types of people have not shown up yet in Mark's story, perhaps they will as the narrative continues. We have to keep listening.[2]

The parable also applies to Jesus himself. He is an itinerant teacher who repeats his message incessantly, his words like seed scattered on inhospitable soil. The more his life and ministry unfold, the dimmer his prospects become. His failure seems complete. Near the end of the story, he cries out in desolation, feeling abandoned by God. His most faithful followers flee in fear; they tell no one of the message that he is alive.

The prophet Daniel has a vision of a humane character, "one like a son of man,"[3] who holds all the nations of the Earth in a loving embrace, in complete harmony with the will of the Creator. In Daniel's vision, Jesus sees the fulfillment of his own mission. In Mark's narrative, Jesus shares a vision with his followers that after the resurrection they will resume his healing work (Mark 13:9–11). From the fertile soil of their renewed faithfulness will spring fields of compassionate action, a great harvest to nourish a broken world.[4]

## THE VALUE OF FAILURE

Some in the crowds hearing this parable are looking for a messiah who will lead a violent overthrow of the Roman Empire. Jesus warns his disciples he is not that person and predicts that they will be "scandalized"[5] by his refusal to lead such an uprising. Mark depicts these disciples—with their bravado, their visions of grandeur, and their impulse to violence—as deeply flawed. They represent the thorn-infested soil, whose brambles will ultimately choke out their professions of faithfulness.

All through Mark's narrative, the Pharisees are like a hard, impermeable path, resistant to the seed. Early on, they collude with the party of Herod to try to get rid of Jesus. The crowds, in contrast, have been his champions. But they turn out to be the shallow growing space, in which the seed sprouts immediately and joyously. There is hope that they will survive the coming heat. But when Jesus goes on trial before Pilate the chief priests manage to turn them against him.

Mark's narrative ends in dramatic and stunning failure, which puts the onus on the listeners to take the story further. Jesus provides the vision of a bountiful harvest of life in harmony with God and nature, springing from seed that falls into even a bit of fertile soil. Has anything taken root? Will his listeners sustain their efforts, as Jesus did, in the face of opposition and failure? Mark's voice falls silent and the story ends.

This parable has universal appeal, especially today. Healing and reconciliation are the work of all people. Religion might be what provokes and sustains some of those who devote their lives to this work. But the kingdom of God of which Jesus speaks is not itself explicitly religious. It is a worldly phenomenon. A *this*-worldly phenomenon.

The idea of a Paradise that is free of corruption and will last forever, as the Creator originally intended, is not some other world to which people will be admitted after death because they have uttered the magical name of "Jesus." Paradise is what sets this Earth back to functioning in such a way that people are no longer sick or insane, starving or consumed by hatred and violence. All the nations of the world and all of its belief systems can participate in this "kingdom," as they generate and are enveloped in what Judy Cannato has described as an expanding field of compassion.[6]

## THOMAS BERRY AND A NEW STORY

It is not only Mark's first-century listeners who are left to step into the silence left by the fleeing women at the end of his story; it is us, his listeners today. In each generation people take up the work of healing the brokenness that causes suffering on Earth. Each generation can do so only if they have a story that accounts for failure as they seek to keep alive their confidence in a bountiful harvest.

In Mark's time the central issue was the Great Revolt against Rome that devastated the land along with the people. Jesus saw this coming and worked to offer Israel an alternative. When his word and deeds fell on inhospitable soil, he kept going, sustained by the prophet Daniel's vision of "one like a son of man." As the Son of Man, he trusted that he would not only rise from death but would come to reconcile the nations, as Daniel prophesied.

War continues to be a source of environmental devastation that brings in its wake hunger, sickness, mental trauma, and death. Nuclear weapons proliferate that have the power to bring death to the entire biosphere. And now a new plotline has been added to the ongoing story of the war and all its consequences: our industrial economy is steadily warming the planet, resulting in rising seas, poisoned water, poisoned air, and poisoned soil, creating human suffering and mass extinctions of vulnerable species.

Thomas Berry's early interest was in world religions and Western history; in the 1970s he became attentive to our moral obligations as a species interacting with all the other species that inhabit our world. He was one of the first religion scholars to warn that human industrial practices contained threats that were rapidly reaching crisis proportions.[7] But the response of most people and public institutions—especially religious institutions—was denial. As in Mark's story, the seed of Berry's message fell mostly on infertile soil.

Berry saw denial as a key element in humanity's war on the planet. We have set ourselves apart from the non-human world, which is seen as having no inherent rights and receives its value only from the way we exploit its resources to meet our needs. Thus we continue to give ourselves permission to impair the integrity of nature, oblivious to the implications of our actions.

The irony, according to Berry, is that our systematic abuse of the natural world has emerged from a culture rooted in Judeo-Christian scripture. As a student of world religions, Berry was in a position to see that it did not arise in the Buddhist, Hindu, Chinese, Japanese, or Islamic worlds. These cultures certainly are not blameless in their treatment of the natural environment, but the damage they have done to the non-human world is not anything on the scale of damage found in the technologically driven, biblically rooted Western Christian culture.

Berry traces this irony to the general discomfort Western culture has exhibited toward nature. The creation myths of Genesis and prophets like Isaiah and Micah make a clear distinction between the Creator and the material world. The creation myths of Genesis make clear that plants and animals were created for the benefit of humans. The creeds, prayers, and catechisms of the Church over the centuries barely mention creation.

Historically, Berry notes, there was a tradition of regarding the Book of Nature as a source of revelation alongside the Book of Scripture. Once scriptures became widely available through the invention of printing, Christians began to regard the written text as the sole source for the knowledge of the divine. As a consequence, most Christians ceased to hear the voices of the world of nature. This widely supported estrangement from the natural world left institutional Christianity generally unconcerned about the emerging threat of species extinction and mute in the face of environmental ruin. Clearly, the situation has improved since Berry's time and some faith groups are now incorporating reverence for the Earth into their beliefs and practices. But the general lack of environmental awareness in the institutional church has made it difficult for the environmental movement to make common cause with the Christian community.

Berry saw the need for a new story of creation that includes what science has learned about the universe and sees the human species as rooted in its material, psychic, and spiritual evolution.[8]

The groundwork for this new story is Berry's understanding of celebrations of the order and beauty of the material world as expressed in the Psalms and the prayers of St. Francis of Assisi, in the cosmology of Thomas Aquinas, and in the cosmological vision of Pierre Teilhard de Chardin. In Aquinas, Berry found a Christian theologian who recognized the cosmological character of every creature. The material world itself indelibly shapes human understanding, and matter itself has a mystical dimension.[9]

From Teilhard de Chardin Berry appropriated the insight that humans are not a species apart from the rest of the universe. The long-standing delusion of human superiority and its license to exploit the material world led seventeenth-century English philosopher John Locke to see nature as "a dull affair, soundless, scentless, colorless: merely the hurrying of material, endlessly, meaninglessly."[10] Teilhard de Chardin recognized that the self-reflective consciousness unique to humans emerged from the way the universe has evolved in the direction of ever-increasing complexity and consciousness. While humans are a differentiated form of life within the cosmos, he believed that the human species "can only be understood through tracing his rise through physics, chemistry, biology, and geology. In other words, man is a cosmic phenomenon, not primarily an aesthetic, moral, or religious one."[11] He was the first to call upon the Church to take seriously what modern science has learned about the evolution of the universe. During his lifetime the seed of his word failed to fall on fertile soil.

Berry went beyond Teilhard de Chardin. He found Teilhard too optimistic about where science and technology could lead us and did not consider its downside. His critique was certainly correct. It was becoming clear in Berry's time that industrial civilization, so confident in itself as an instrument of progress, was beginning to move the world along the path of environmental destruction, all the way to ecocide.

How can we reconcile the old, life-destroying creation story with a new, life-affirming one? Berry advocates an accessible narrative that blends the insights of science with the inspiration of spirituality, and a call to action that makes a real difference in the world. It needs to be a story the nonscientist can understand.[12] He would say to his students, "With a story . . . people can endure catastrophe. And with a story they can gather the energies to change their lot."[13]

An important motif in this new narrative is that our unique capacities as humans need not intrude on the universe or impair it; they emerge in humans because these capacities were present in the universe from its very beginning. "In ourselves the universe is revealed to itself as we are revealed to the universe. . . . [E]very being in the universe articulates some special quality of the universe in its entirety."[14]

While it might first appear that Jesus's vision of the harvest in Mark and Berry's vision of an emerging era of ecological consciousness are different, a bit of reflection reveals their deep connection. War and environmental devastation are inextricably intertwined in Israel's scriptures. Jeremiah 50:3 says this quite clearly:

> A nation from the north advances against it,
> making the land desolate
> So that no one can live there;
> human beings and animals have fled.

Jesus's vision of the tribulation that would precede the coming of the Son of Man (Mark 13) included war, persecution, and destruction of the natural world: the sun and the moon would be darkened; stars would fall from the heavens. This image of cosmic collapse is drawn from Israel's prophetic tradition. This combination of war and the collapse of nature is ubiquitous—a fact often ignored when the media report on war. What is new in human history, as Thomas Berry was one of the first to see, is that our industrial practices now threaten to devastate the entire ecosystem.

The Hebrew prophets envisioned both the healing of the nations and the healing of the natural environment. Ezekiel 36:33–36 is an especially striking example of this:

> Thus says the Lord GOD: When I cleanse you of all your guilt, I will resettle the cities and the ruins will be rebuilt. The desolate land will be tilled—once a wasteland in the eyes of every passerby. They will say, "This once-desolate land has become like the garden of Eden. The cities once ruined, laid waste and destroyed, are now resettled and fortified." Then the surrounding nations that remain shall know that I, the LORD, have rebuilt what was destroyed and replanted what was desolate. I, the LORD, have spoken: I will do it.[15]

Ecofeminists such as Ivone Gebara have observed that the issues of war, our disastrous abuse of the earth, and the historic domination and abuse of women by men are deeply connected. In Mark's narrative the connections among these three issues do not appear on the surface. In a review of the two world wars, Vietnam, and the Rwandan war, Gebara describes the connections well:

> In any act of aggression against nature, the real target is the local human inhabitants. So there is no direct intention to destroy the fauna, flora, or fish in the rivers. Despite this, however, aggression against human beings also becomes aggression against all of nature. . . . We poison nature in order to destroy all of life because in that way we destroy human life. . . . We use nature as a victim and also as a weapon.
>
> The same happens with women. They too are a war target, and their bodies are used as means of sowing terror in the civilian population. They are raped and brutally beaten, and thus used as pawns in the process of provoking even greater hatred among resistance groups. . . .
>
> We do not often carry out this sort of historical analysis. The use of nature . . . and of women as weapons in war and conquest gets little attention in our analyses. . . .
>
> We usually count the dead in war, but we almost never mention the destruction of the environment, the death of animals, the poisoning of natural springs, and the destruction of the present and future means of survival of those who have not died.[16]

Add to this Elizabeth Johnson's observation that "the ecological crisis does not get to the heart of the matter until it sees the connection between exploitation of the earth and the sexist definition and treatment of women."[17] "The ruling man's hierarchy over women and slaves extends also to nature, most often symbolized as female. . . . [B]oth women and the natural world are separated from the men they bring forth and sustain. Both are assigned instrumental value, with little or no intrinsic worth apart from their potential to serve the needs and desires of men."[18]

In hearing Jesus share his vision of the coming of the Son of Man, it is important for us to be as familiar with the vision in Daniel 7 as Mark's listeners likely were. The book of Daniel was a popular focus of conversation and interpretation among Israelites longing for the redemption of their nation.[19] In Daniel's vision, the "one like a son of man" unites the nations under a humane as opposed to a bestial rule. But it would be a mistake to think that Jesus's vision for the future harvest was limited to the reconciliation of the nations. It also includes healing the relationship between humans and nature, between men and women, a commitment to nonviolence, and the reconciliation of enemies within the tribe of Israel and between Israel and the Gentiles. These themes in Mark's story might provoke the listeners to wonder at the connections between these three kinds of brokenness in the world.

## A NEW STORY OF THE UNIVERSE

Jesus's own interpretation of the parable in Mark 4:14–20 treats the different soils as metaphors for different responses to the seed, which is his word or message. The new story of the universe as told by modern science encourages us to reconsider the symbolism of the various environments Jesus describes. There is another way to think about it. As a species we are embedded in the very soil of the Earth, perceived already by the Genesis narrator and made even more clear by the modern story of the evolution of the universe. Our capacity for making a free, self-reflective response to heal a wounded world is ours by virtue of our own evolution from the physical Earth. How shall we behave toward our fellow humans and the other species with whom we share the same soil, water, and air as the source of our life? Shall we continue to abuse the Earth on which we are utterly dependent for sustenance? Will males continue to abuse the women who conceive our offspring in their soil-dependent bodies and nurse them at their breasts? Shall we differentiate among other humans and go to war against them? Jesus's mission to restore the Creator's rule to Earth involves healing broken relationships with our fellow humans—and with the species that share our planetary home. Only when we recognize our deep connectedness with every creature will we evolve into a larger, more complex, more beautiful entity for which our self-reflective freedom equips us.

When Thomas Berry's *The Dream of the Earth* was published by the Sierra Club in 1988, it was considered a landmark in the literature of the ecological movement. Some of its most compelling passages articulate the challenges that lie ahead, particularly for young people. Berry thought deeply about how institutions of higher learning would equip future generations for the task of rescuing and restoring the Earth. He wrote of the need for new language to describe what it means to be human and for ways that the human species can live in harmony with all the other species. He understood that the transition from a scientific-technological period to an ecological period in our evolution will be turbulent. Berry encourages universities to help their students to embrace a spirit of adventure as they participate "in one of the most significant ventures ever to take place in the entire history of the planet." Their challenge is to see their role to be "creating a future worthy of that larger universal community of beings out of which the human component emerged and in which the human community finds its proper fulfilment." Berry put it this way:

> Here we might observe that the basic mood of the future might well be one of confidence in the continuing revelation that takes place in and through the earth. If the dynamics of the universe from the beginning shaped the course of the heavens, lighted the sun, and formed the earth, if this same dynamism brought forth the continents and seas and atmosphere, if it awakened life in the primordial cell and then brought into being the unnumbered variety of living beings, and finally brought us into being and guided us safely through the turbulent centuries, there is reason to believe that this same guiding process is precisely what has awakened in us our present understanding of ourselves and our relation to this stupendous process. Sensitized to such guidance from the very structure and functioning of the universe, we can have confidence in the future that awaits the human venture.[20]

We have reason to hope that the process that has brought the universe to its present state of astonishingly complex development will guide us into our evolving future. But how can we be confident with the knowledge that we are minuscule specks of self-conscious and reflective awareness in a world 13.7 billion years in the making, the inhabitants of one small planet amid uncounted galaxies and billions of stars?

## SMALL BEGINNINGS, BIG IMPACTS

How can humans move creation to renewed health? One insight from modern science resonates with the narratives of Judeo-Christian scripture: If everything in the universe is connected and even a small change can affect the entire system, we can imagine having an impact.

In the Hebrew scriptures, sending the flood was a strategy of throwing an enormous solution at an enormous problem. The call of Abraham and Sarah constitutes the opposite approach. This time God started with just two people—two people who were past their prime. They were too old to bear a child. But God called them to be a new start in transforming the world:

> The LORD appeared to Abraham by the oak of Mamre, as he sat in the entrance of his tent, while the day was growing hot. Looking up, he saw three men standing near him. When he saw them, he ran from the entrance of the tent to greet them; and bowing to the ground, he said: "Sir, if it please you, do not go on past your servant. Let some water be brought, that you may bathe your feet, and then rest under the tree. Now that you have come to your servant, let me bring you a little food, that you may refresh yourselves; and afterward you may go on your way." "Very well," they replied, "do as you have said."
>
> Abraham hurried into the tent to Sarah and said, "Quick, three measures of bran flour! Knead it and make bread." He ran to the herd, picked out a tender, choice calf, and gave it to a servant, who quickly prepared it. Then he got some curds and milk, as well as the calf that had been prepared, and set these before them, waiting on them under the tree while they ate.
>
> "Where is your wife Sarah?" they asked him. "There in the tent," he replied. One of them said, "I will return to you about this time next year, and Sarah will then have a son." Sarah was listening at the entrance of the tent, just behind him.

> Now Abraham and Sarah were old, advanced in years, and Sarah had stopped having her menstrual periods. So Sarah laughed to herself and said, "Now that I am worn out and my husband is old, am I still to have sexual pleasure?" But the LORD said to Abraham: "Why did Sarah laugh and say, 'Will I really bear a child, old as I am?' Is anything too marvelous for the LORD to do? At the appointed time, about this time next year, I will return to you, and Sarah will have a son." Sarah lied, saying, "I did not laugh," because she was afraid. But he said, "Yes, you did." (Genesis 18:1–15)

From that point on, Israel tells stories of how time after time God chooses to work through those who, by the world's standards, were weak and unqualified. The nation of Israel itself is a prime example: Who were the Israelites when God formed them into a nation? A bunch of escaped slaves.

In his book *Jesus of Nazareth: What He Wanted, Who He Was*, New Testament scholar Gerhard Lohfink coins the term "The Abraham Principle."[21] Lohfink points out that God's sending of Jesus is the climactic instance of the strategy that began in choosing Sarah and Abraham. Who is Jesus? A peasant from Nazareth. He does some powerful and astonishing things. But he is not especially successful in winning people to his cause. In fact, his own work ends in abject failure. By what means will it move forward, if it does? By the willingness of people just like him to take up his work. More often than not, they will fail as much as Jesus did.

And yet both Jesus and many who have come after him continue to do what they can, keeping the world afloat on a sea of compassion. The outcome of those efforts has yet to be seen.[22] Those who continue to pursue the goal do so in the belief that even small efforts make a difference. The little they do creates a field of compassion that merges with other fields of compassion. Even the little they do impacts the whole system. Everything in the universe is bound together. Even something as seemingly minor as the flapping of a butterfly wing can change the weather pattern on the other side of the Earth.

## GOD'S FARMING TECHNIQUE

There is one more element to consider about Jesus's metaphor of the different soils. In the parable, the farmer does his work according to the custom in Jesus's time, when little technology existed. He broadcasts the seed widely by flinging it with his hand.

Jesus's parables often involve some note of exaggeration, even of absurdity. We may have an instance of that in this story. Did Palestinian farmers in Jesus's time really fling the seed onto a beaten-down path, onto rocky ground, or into briar patches? From the vantage point of mechanized modern farming the whole process of sowing seed in this way may seem strange. But even in Jesus's day people may have found this indiscriminate casting of the seed on such poor soil outrageous. The British biblical scholar C. H. Dodd, in a landmark 1935 book, *The Parables of the Kingdom*, offered a provocative description of how absurdity functions as a rhetorical device:

> At its simplest the parable is a metaphor or simile drawn from nature or common life, arresting the hearer by its vividness or strangeness, and leaving the mind in sufficient doubt about its precise application to tease it into active thought.[23]

If we reflect on this in relation to Mark's entire story of Jesus, our minds are teased into thought, not only by Jesus's explanation of his parable to his disciples out of earshot of the crowd, but by this odd note of exaggeration. Various characters disappoint us by looking like the hardened path, rocky and shallow soil, or thorn-infested ground. Are we not provoked to wonder why any farmer with his head screwed on right would scatter seed on such soils in the first place?

There is no definitive response to that question. But it seems to reflect the inclusive goal and therefore inclusive strategy of Jesus's mission. He recruits helpers for his work by calling them away from their daily tasks. He asks for no resume or qualifications. He does not embark on his work in Galilee with an eye to identifying the most faithful people. He calls the least faithful, the sinners, and even tax collectors. He goes not just to the children of Israel but to the aggressive outsiders of Syrophoenicia. He broadcasts the seeds of his words and deeds without discrimination.

Where will they bear a great and bountiful harvest? Jesus does not know. Perhaps it is for that reason he excludes no one. And here lies a lesson as we contemplate the enormous task of reversing the curse of environmental degradation and following our collective destiny as participants in the evolutionary process: If some of the soil on which the seeds of our activism fall turns out to be rich and nurturing, the abundance of the harvest will be beyond belief.

# 8

# Here and Now

For most of Christian history the default version of how humankind engages with God went something like this: Jesus came to Earth and died for my sins so that after I die, my soul can go to heaven and be with God forever.

The idea of leaving a sullied world for the bliss of the afterlife continues to be a comforting notion to Christians today. But scripture scholars, theologians, and liturgists are shifting the emphasis of the way we tell the story of humankind's quest for connection with God from the future to the present: In the present, in partnership with Jesus, we are challenged to deliver the Earth and its creatures from brokenness, violence, and death.[1]

Mark's story of the life of Jesus recounts his public ministry as a series of transformative acts of healing and compassion: curing the sick, casting out demons, feeding the hungry, calming a storm, welcoming women, gathering outcasts. He teaches in parables rich in metaphors of nature, farming, and domestic life. He engages in respectful controversy with those who object to what he is doing. He leads his disciples on a journey to the center of Israel's religious and political life, instructing them and those he meets along the way about money, sex, and power. Only the final two chapters focus on his suffering, crucifixion, and resurrection. Taken together, these elements of Mark's story complete Israel's long narrative of salvation, in which a fully human child of God repairs the damage done to life on Earth when Adam and Eve reject God's partnership. Mark sees Jesus as the one who lifts the curse on humankind that the narrator of Genesis describes.

But over the centuries, Jesus's work of renewing and healing all creation has played only a minor role in the development of Christian doctrine and liturgy. In his 2010 book *The Four: A Survey of the Gospels*, Protestant theologian Peter Leithart makes this observation:

> Classical Christology has its distortions. It has pitched its tent almost exclusively at the margins of the gospel story. The few narratives of Jesus' birth, along with John's great prologue, have been central to discussions of the nature of the Incarnation; the stories of the crucifixion have played a central role in the development of atonement theologies, though not nearly so great a role as Paul's discussions of the death of Jesus. The period between birth and death, the life and ministry and miracles of Jesus, have played very little role in the development of Christology.[2]

Leithart has a point. The second-century Apostles' Creed, affirmed by nearly all traditional Christian denominations, summarizes the story of Jesus in this way:

> Who was conceived by the Holy Ghost,
> Born of the Virgin Mary,
> Suffered under Pontius Pilate,
> Was crucified, dead, and buried:
> He descended into hell;
> The third day he rose again from the dead:
> He ascended into heaven,
> And sitteth at the right hand of God the Father Almighty:
> From thence he shall come to judge the quick and the dead. (*The Book of Common Prayer*, 1945)

There are no hints here about Jesus's restored harmony with nature, or his work of healing, nurturing, and reconciling those in distress. Nor is there anything about God's saving work in Israel. The creed begins with a statement of belief in God the Father, and then moves immediately to "Jesus Christ, his only Son, our Lord." It proceeds directly from Jesus's birth to his death.

The same is true of the summaries of the story of Jesus heard by Christians in the Eucharistic Prayers. Here is the portion of Eucharistic Prayer II from the Roman Mass:

> It is truly right and just, our duty and our salvation, always and everywhere to give you thanks, Father most holy, through your beloved Son, Jesus Christ,
> your Word through whom you made all things,
> whom you sent as our Savior and Redeemer,
> incarnate by the Holy Spirit and born of the Virgin.
> Fulfilling your will and gaining for you a holy people,
> he stretched out his hands as he endured his Passion,
> so as to break the bonds of death and manifest the resurrection. (*The Roman Missal*, 2011)

Note again the absence from this celebration of the life of Jesus of any mention of his compassionate actions prior to his death.

## CHANGING PERSPECTIVES

How did this view of the story of Jesus become the default narrative of the Church? Part of the answer resides in the power that Greek philosophy and its ideas of the afterlife had over both Israelite and Gentile Christians as they listened to the gospel narratives.[3]

Three hundred years before Jesus of Nazareth walked the Earth, the Macedonian warrior known as Alexander the Great conquered a swath of the civilized world stretching from the lands of the eastern Mediterranean to India. He pursued a policy of deliberately spreading the culture of Greece—its language, its philosophies, its institutions, such as the market, temples, theater, and gymnasium. Historians label this cultural tsunami "Hellenism" after the Greek word for Greece, *Hellas*.

Every culture absorbs new information in terms that it can understand. The question is always how much of the original message is retained and how much has changed. Most listeners adjust the meaning of what they are hearing without being at all conscious of what they are doing. It is the work of scholars to notice and account for these changes.

In the time of Jesus, no one could escape the impact of Greek civilization any more than people and nations today can escape the impact of Western globalization. Hellenistic culture affected all Israelites, one way or another. Very early on, Gentile followers of Jesus became increasingly influential in the Church. Both they and Hellenized Israelites who had joined the Christian movement strove to understand the story of Jesus from the perspective of the culture in which they had been raised. And their culture was strongly influenced by the philosophy of Plato.

The Platonic tradition led many early Christians to think of the human soul as immortal, a spiritual substance that exists beyond the body. The human predicament is that an embodied soul dwelling on Earth needs to be restored to its proper place—free of the physical body—in the afterlife.[4] In contrast, those who held to Israel's story of the cosmos believed that the human predicament was an Earth wounded by human rebellion against the Earth's creator. If the soul of an individual survived the death of the body, it would be resurrected to a transformed body in a new heaven and new Earth.

When Gentiles and Israelites educated in the developing Platonic tradition joined the Jesus sect, they brought their cultural worldview with them, interpreting the story of Jesus from Plato's point of view. Some sought secret knowledge as the way to escape the soul's imprisonment in the material world. Scholars call them "Gnostics," from the Greek word for knowledge. They rejected Israel's scriptures, which spoke of the material world as good. Scholars describe other converts to the Jesus sect as "orthodox" or "proto-orthodox." Unlike the Gnostic Christians, these believers continued to engage with Israel's scriptures and adapted Israel's perspective to a Platonic view of salvation as they interpreted the story of Jesus.

Philo, a first-century Israelite philosopher who advocated a Platonic interpretation of Israel's scriptures, had a significant impact on the thinking of influential Gentile Christians living in Alexandria in the second and third centuries. He was deeply influenced by Plato's view of the relationship of the soul to the body.[5]

This view of salvation became dominant as Gentile believers strongly influenced by Hellenism began to flood into the Church. It resulted from a natural, even unconscious interpretation of the story of Jesus from the Hellenistic perspective that was very different from Israel's ideas about the body and soul.

Most Christians believe in life after death, and they picture it as the life of their disembodied soul in the spiritual realm called heaven. In this view the Earth plays no further role in their life.[6] Conclusion: Christianity is an "otherworldly religion."[7]

How could Christians listen to the Gospel of Mark and hear in it a Platonic view of salvation? The simple answer is that Mark's minimalist storytelling can often be understood in more than one way. The same phrase that means one thing to an Israelite signifies quite another thing to a Gentile. Mark 10:17 contains an example: A man runs up to Jesus, falls on his knees in front of him, and blurts out, "Good Teacher, what must I do to inherit eternal life?" What was he asking? From a Platonic point of view, he was asking how his soul could attain eternal life in heaven. From an Israelite point of view, he was asking how he could live so that his soul would be clothed in a new body to live eternally on a transformed Earth.

## EVOLVING INTERPRETATIONS

As Christianity grew and evolved over the centuries, other forces continued to shape the faith. Massive cultural changes in Western civilization have been under way since the fourteenth century, when the Italian Renaissance energized our evolving understanding of the physical world and the capacity of human beings to shape our lives and our environment.

Artists began to paint scenes from scripture in which biblical characters were presented in contemporary settings and which emphasized the natural environment. These artists recovered from ancient Greece and Rome the practice of depicting the human body realistically instead of in the stylized ways of more iconographic painting that prevailed in the early church. Leonardo da Vinci even secretly dissected human cadavers in order to paint and sculpt the human figure more accurately.

The seventeenth century saw the birth of the Enlightenment and the beginnings of modern science, through which human beings have been reshaping our environment in ever more radical ways. Our human impact on our natural environment is reaching a critical stage that people are slowly but reluctantly beginning to recognize.

Enlightenment thinkers like Thomas Jefferson were taking a new interest in the human Jesus and what he had to offer for the living of our daily life. As a humanist, Jefferson was put off by the traditional Church's focus on supernatural ideas like the divinity of Jesus, his Virgin Birth, miracles, the Trinity, and the resurrection of the body. These ideas offended reason and common sense and were of no practical value. Jefferson was interested in what Jesus taught about ethical living.[8]

In the mid-twentieth century, in the wake of two world wars, new concerns emerged for human freedom and social justice. Nation after nation gained independence from imperial rule. India was a prime example: Under the leadership of Mohandas K. Gandhi, independence was initially achieved through a strategy of nonviolence. The State of Israel was created in the wake of the genocidal efforts of the Nazis to eradicate the Jews. Martin Luther King Jr. successfully led protests against racial segregation in the United States, again pursuing a strategy of nonviolence. Since that time we have seen black South Africans succeed in another nonviolent revolution under the leadership of Nelson Mandela. But these are bright spots in an otherwise bleak landscape.

Today, Gandhi's ideas of nonviolence have been eclipsed by the aggression of populist nativists who demonize the waves of migrants fleeing war, hatred, and natural disasters. In the United States, a gun culture proliferates and mass shootings have become commonplace. The postwar ethos of cooperation among nations is giving way to authoritarian leaders, a hard-hearted isolationism, and the demonization of the poor, the homeless, and the migrants and strangers among us.

The strategy of nonviolence in the service of justice for oppressed peoples and a growing concern for human impact on our natural environment have combined with developments in recent biblical scholarship to lead liturgical scholars to develop resources for Christian worship that reflect and speak to these cultural changes. They reflect a greater concern for how we are to participate in the saving of the world in which we now live.

## THE DEFAULT HEARING OF MARK'S STORY

But these new insights into the biblical vision for the future of the Earth exist alongside the Church's default theology. The idea that Christianity should focus more on the afterlife than on this life continues to influence the Church and the culture. It continues to govern the way Christians think and speak. The power of that default understanding of the story of Jesus is reflected in the reluctance of many Christian congregations to leave behind the earlier formulations of the Eucharistic Prayer and the traditional creeds for new ones that incorporate the story of God's saving work in Israel, in Jesus's compassionate ministry to the poor and the social outcasts, and the ultimate healing of the Earth.[9]

Here, for example, is Eucharistic Prayer IV from *The Roman Missal*, English Translation according to the Third Typical Edition (657):

> We give you praise, Father most holy,
> for you are great and you have fashioned all your works in wisdom and in love.
> You formed man in your own image and entrusted the whole world to his care,
> so that in serving you alone, the Creator, he might have dominion over all creatures.
>
> And when through disobedience he had lost your friendship,
> you did not abandon him to the domain of death.
> For you came in mercy to the aid of all, so that those who seek might find you.
> Time and again you offered them covenants
> and through the prophets taught them to look forward to salvation.
>
> And you so loved the world, Father most holy,
> that in the fullness of time you sent your Only Begotten Son to be our Savior.
> Made incarnate by the Holy Spirit and born of the Virgin Mary,
> he shared our human nature in all things but sin.
> To the poor he proclaimed the good news of salvation,
> to prisoners, freedom,
> and to the sorrowful of heart, joy.
>
> To accomplish your plan, he gave himself up to death,
> and, rising from the dead, he destroyed death and restored life.
> And that we might live no longer for ourselves
> but for him who died and rose again for us,
> he sent the Holy Spirit from you, Father, as the first fruits for those who believe,
> so that, bringing to perfection his work in the world, he might sanctify creation to the full.

When people are offered a choice of creeds or Eucharistic Prayers, they often say that they prefer the older forms. This preference for the older forms that omit these parts of the Biblical story, so essential to Israel's perspective, may have a lot to do with the fact that so few Christians or official Church bodies speak out in favor of government policies that care for the poor instead of the wealthy, limit the havoc industry creates in nature, and put public money and imagination to work in pursuit of nonviolent ways of confronting our enemies.[10]

## SALVATION IN THE TWENTY-FIRST CENTURY

Our capacity to dominate the Earth has led to an ever-greater loss of control, and the continued life of our planet is threatened by the very technology we have developed.

In his 2006 book, *Eden's Garden: Rethinking Sin and Evil in an Era of Scientific Promise*, Richard J. Coleman observes that modern science has been naïve about the question of good and evil. The assumption is that if we *can* do something technologically we *should* do it. The only principle that seems to regulate technological development is whether there is someone around who can make money from developing it.

As Pope Francis makes clear in *Laudato Si'*, his encyclical on the environment, the current crisis is a spiritual and moral problem, not a strictly technological one. The question before us is articulated in the foundational myth of the Hebrew Bible, the story of Adam and Eve: Will humanity, created of dust, use its God-given free will to make responsible moral choices about the Earth? Will our species resist the temptation to take what delights the eyes and promises to make us wise long enough to ask whether doing so is good for us and good for the world into which we have been born?

When the story of Jesus entered the Gentile world, motifs in the narrative were adjusted to conform to the Platonic worldview. In the same way, the story of the tree of the knowledge of good and evil requires new interpretation but remains critical in the context of modern science's story of evolution.

In *Freedom, Suffering and Love*, Andrew Elphinstone brings clarity to our current challenge. Genesis pictures a time of human innocence. In contrast, science yields a story of how over the course of human history our knowledge of good and evil has evolved. We began in fear of the other; to be good meant to be aggressive and violent and to dominate others in order to survive against threat. In our time, the human tendency to violence threatens not only our own species but the life of every other species. What is called for now is love, that newcomer in the history of evolution that goes against our most primitive instinct of fear. We desperately need to choose which is evil and which is good: violence or the kind of love embodied in Jesus of Nazareth.

The default reading of Christianity says that this world can never be anything other than a realm of corruption and death. Mark's story challenges that notion. Rather than coming to Earth to die so that we might go to heaven, Jesus came to transform this world into a better place, in which we no longer give in to the temptations that impair both humankind and nature itself.

What is our ultimate vision for the Earth? Where is the relationship of this planet and its inhabitants in the unfolding cosmos? The prophet Isaiah envisioned a transformation of both divided humanity and divided nature. Isaiah 2:2–4 pictures the nations streaming to Mount Zion to join Israel in the worship of God, after which they will beat their swords into plowshares, their spears into pruning hooks and abandon war. Isaiah 11:6–9 imagines the wolf lying down with the lamb and the children playing next to the dwelling places of snakes. Other Hebrew prophets share both of these visions. Hosea's vision in 2:16–22 is another stunning example. In verse 20 a new covenant with Israel will extend to nature and rid the world of war:

> I will make a covenant for them on that day,
>   with the wild animals,
> With the birds of the air,
>   and with the things that crawl on the ground.
> Bow and sword and warfare
>   I will destroy from the land,
>   and I will give them rest in safety.

The prophets' vision of an Earth transformed may be deemed unrealistic, but it was the same vision that led Jesus of Nazareth to believe he could make a difference. And many people today long to do what they can to transform life in the present. The critical factor in Isaiah's vision is his belief that the Earth and the universe exist because there was a time before it existed. In his book, *We Make the World by Walking*, Evangelical clergyman Brian McLaren reflects on what brought our universe into existence:

> Imagine a creativity, brilliance, fertility, delight, energy, power, glory, wisdom, wonder, greatness, and goodness sufficient to express itself in what we know as the universe. . . . [T]ry to imagine that this is also the great, big, beautiful, mysterious goodness, wholeness, and aliveness into which all of us and all creation will be taken up.[11]

Israel had such a hope. Jesus shared that hope. So did Mark. Rather than coming to Earth to die so that we might go to heaven, Mark's story shows how Jesus came to transform this world into a better place, in which humans no longer yield day after day to the temptations that impair humankind and nature itself.

In harmony with the Creator, Jesus devoted himself to deeds that created a field of compassion that would reverse the curses catalogued in Genesis 3:14–19. He relinquished his life rather than give fear and hatred one more victory.

We have shown in these pages how small actions inspired by the life and teachings of Jesus can have impacts that extend far beyond time and space. Science has shown us that everything in the universe is connected, so that an event as small as the flapping of the wings of a butterfly on one side of the Earth can affect conditions thousands of miles away. The cumulative effect of individual acts of compassion can heal and transform the world.

Jesus's vision has yet to be realized. Will it ever come to pass? That depends on how we choose to live our lives. This is a spiritual and a moral question. Will we dismiss Jesus, Mark, the visionaries, prophets, and saints as impractical dreamers and continue to use our intelligence to destroy ourselves and all that we love? Or will we trust that the creative power that brought this world into being can transform it?

We can begin by committing ourselves to starting small. Perhaps that will be enough.

# Afterword
## Echoes and Allusions

In tradition and in life, there is more to most music than a melody. There are harmonies, echoes of other melodies, and melodies within melodies. In the same way, there is always more to a story than getting from a problem to the resolution. In tradition and in life, some stories often contain echoes of other stories. They meet and diversify, harmonize and complete each other, echo and evolve just as the material universe evolves around us. In scriptural study, these echoes are what scholars call "allusions." But listeners cannot know they are hearing echoes if they have not heard the original song. They cannot recognize allusions until they know the characters, plotlines, metaphors, or symbolism of earlier stories.

Does Mark's story of Jesus allude to the story of Adam and Eve in Genesis 2 and 3? More particularly, are there allusions in Mark to the Genesis story of a Paradise that has been cursed? Intentional allusions cannot be identified with total certainty. The echoes between Genesis and Mark's gospel may be nothing more than similar but coincidental sounds. How confident can we be that Mark meant for his listeners to recognize them?

In *The Conversion of the Imagination: Paul as Interpreter of Israel's Scripture*, New Testament scholar Richard B. Hays proposes seven criteria for identifying scriptural allusions: *availability, historical plausibility, recurrence or clustering, volume, thematic coherence, satisfaction,* and *history of interpretation*. When there is a cumulative satisfaction of the seven criteria, it is likely that an allusion is present. Hays is attuned to the ways in which the New Testament echoes the storytelling elements of Israel's traditions. His criteria provide clues to the presence of the curse as a master allusion in Mark.

The story of God's curse was certainly *available* to Israelites like Mark. It was a major storyline, placed at the beginning of Israel's scriptures, setting the scene and the tone of narratives to come. Its depiction of an original harmony disrupted by the free decision of the first humans to turn against God's will is foundational to the Israelites' story of themselves, the existential dilemma they strove to understand and resolve. Two works popular in Mark's day clearly refer to it: The Book of Wisdom 1:12–2:24 in the Greek canon and 1 Enoch 69:6, which was widely circulated apart from the canon.

It is *historically plausible* that Mark was conscious of the curse storyline as he spun his narrative about Jesus. In his reply to the Pharisees' question about divorce in Mark 10:6–8, Jesus explicitly quotes Genesis 1:27 and 2:24.

Mark's story of Jesus's temptation contains a *cluster* of narrative elements that cohere with themes in Genesis 1–3: Jesus's harmony with the wild beasts, his successful resistance of temptation, and his subsequent announcement of the inauguration of God's rule. Together these raise the question whether Jesus is a New Adam who will make it possible for humanity to turn back to God. The narratives rarely echo each other in specific words; the word for "wild beasts" is an exception. But the themes of temptation and obedience resonate clearly. Consequently, Hays would say the *volume* is low, but the cluster of narrative motifs presses the allusion on the attentive listener.

Many Christians see the allusion to Adam and Eve's temptation by the snake in Jesus's temptation by Satan. It seems a natural thing to do. But was Mark the only Israelite to correlate the snake with Satan? *Historical plausibility* increases when we see that other Israelites made this connection, too. The composer of the Book of Wisdom in the Greek canon says, "But by the envy of the devil, death entered the world" (2:24). So does a first-century CE composition called the "Life of Adam and Eve." Eve retells the story of the fall to her children by Adam: "And the devil spoke to the serpent . . . Why do you eat of the weeds of Adam and not of the fruit of Paradise? Rise and come and let us make him to be cast out of Paradise through his wife. . . . And immediately he suspended himself from the walls of Paradise. . . . And he said to me, 'What are you doing in Paradise?' I replied, 'God placed us to guard it and eat from it.' *The devil answered me through the mouth of the serpent*" (Apocalypse 15–17, trans. M. D.

Johnson, in *The Old Testament Pseudepigrapha*, Vol. 2, ed. James H. Charlesworth [Garden City: Doubleday, 1985], 277–279). So Mark is not the only one to make this connection.

And what of the presence of angels? Commenting on the story of Jesus's temptation in his commentary on Mark, Joel Marcus reports that the much later Babylonian Talmud, Sanhedrin 59b, also records a Jewish legend according to which ministering angels prepare food and drink for Adam in Eden.

It is *historically plausible* that an Israelite narrator would tell a story about the reversal of the curse. Scholars have identified this as a theme in other Israelite compositions, like the Song of Songs and Paul's Letter to the Romans.

An allusion to the loss of Paradise in Mark's story of Jesus's temptation establishes a *theme that gives coherence* to Mark's narrative and provides *satisfaction*. The comprehensive theme of Mark's narrative is to draw out the story of Jesus's work of inaugurating the reign of God (Mark 1:15). Would a listener not expect that when God's rule returns, things would begin to happen that would heal the various forms of brokenness symbolized in the curse?

When viewed in relation to the curse, the contours of Jesus's words and deeds begin to stand out more clearly. In the immediate wake of resisting Satan, a clear picture emerges: intimations of interest in healing and the reestablishment of harmony between humans and the forces of nature. It begins with Jesus's companionship with the wild beasts and continues with the calming of wind and waves.

Mark's narrative also addresses the curse of male domination, as women challenge Israel's patriarchal norms by initiating encounters with Jesus. He welcomes them rather than rebuking them. Toward the end of the story Mark tells us that many women had been part of his company of followers in Galilee, exemplifying the service he commends to his male disciples and playing the role of sisters in his spiritual family (3:34; 10:30).

Jesus's struggle against violence and death also acquires a more crucial importance when we see that any acceding to those strategies in order to establish the reign of God would perpetuate the curse of death.

Hays's remaining criterion is the *history of interpretation*. This refers not to interpretation in the works of other Israelites contemporary to Mark but to interpretation in the subsequent life of the Church. Hays's remarks about the later interpretation of Paul's letters align with this book's observation about how orthodox Christianity has understood—or failed to understand—the Gospel of

Mark: "The Christian tradition early on lost its vital connection with the Jewish interpretive matrix in which Paul lived and moved; consequently, later Christian interpreters missed some of Paul's basic concerns" (43). When we again hear in Mark's gospel the echoes of Israel's story of Paradise and Paradise Lost, we hear new harmonies to which later Christian tradition had become deaf.

Hays goes on to ask why the new understanding we are unearthing has not been made more clear by the ancient author. Why the ancient speaker's reticence? Hays's answer in relation to Paul articulates what we have been saying in this book about Mark's style as a narrator. He writes that Paul "hints and whispers all around Isaiah 53 but never mentions the prophetic typology that would supremely integrate his interpretation of Christ and Israel. The result is . . . Paul's transumptive silence cries out for the reader to complete the trope" (44). Hays further appeals to us to "give Paul and his readers credit for being at least as sophisticated and nuanced in their reading of Scripture as we are. Everything about Paul's use of OT texts suggests that his 'implied reader' not only knows Scripture but also appreciates allusive subtlety" (49).

In this book, we appeal to our readers to give Mark and his listeners the same credit.

But we add another layer to our modern listeners' response to Mark. The harmonies of Israel out of which Mark's melody arises resonate in the modern listener with our own concerns: the struggles of the weak and the marginalized, especially the sick and the traumatized; the women who provide the most basic necessities of life yet are virtually invisible to those in power; the strangers who are not welcomed in our land and turned away with cruelty; the Earth itself, the source of our very life, yet so severely abused.

Scientific advances continue to reveal the wonders of the natural world whose glory the ancient Israelites already praised. As this chorus of discovery begins to crescendo, it invites us to participate in the great universal harmony.

As we discern more clearly voices other than our own, we may find the courage to conquer our tribalisms, relinquish our fears, and join together in the music that the universe has been singing for 13.7 billion years. What a wonder that in this ancient symphony the voice of humanity has been given a turn. Like Adam and Eve in Paradise, harmony or cacophony is ours to choose. Let us hope that in our evolving awareness of the sanctity of all creation and the power of compassion to heal a wounded world, the beauty of our collective voices will only increase.

# Notes

## INTRODUCTION

1. Catherine Brown Tkacz, *Αληθεια Ελληνικη: The Authority of the Greek Old Testament* (Etna, CA: Center for Traditionalist Orthodox Studies, 2001), argues that many of the differences between the Greek and Hebrew canons were introduced into the Hebrew canon in the early Christian era in response to Christian interpretation. Early Christians interpreted both single words and entire narratives about women present in the Greek canon to show women as types of Christ. The absence of their Hebrew equivalents in the Masoretic text suggests deliberate omission by the rabbis during the early centuries of the Christian era in order to distance themselves from the Christian use of Israel's Greek scriptures.

2. For recent scholarship on this development of the two sects called Christianity and Judaism today, see Daniel Boyarin's *Border-Lines: The Partition of Judaeo-Christianity* (Philadelphia: University of Pennsylvania, 2004) and his article "Semantic Differences; or, 'Judaism'/ 'Christianity,'" in Adam H. Becker and Annette Yoshiki Reed, eds., *The Ways That Never Parted: Jews and Christians in Late Antiquity and the Early Middle Ages* (Minneapolis: Fortress, 2007).

3. Joel Marcus, *Mark 1–8: A New Translation with Introduction and Commentary* (New Haven, CT: Yale University Press, 2000), 490. In his comment on Mark 8:1–9, Marcus says: "We have seen that the biblical allusions in 7:37, which are to Gen 1:31 and Isa 35:5–6, suggest that Jesus' healings are part of a creative feat comparable to the primordial act of creation and that they signal the advent of God's new age."

Richard Bauckham, *The Bible and Ecology* (Waco, TX: Baylor University Press, 2010), 127–129, sees in Mark's temptation a story that prepares Jesus for his preaching about "the return to Eden we find in Isaiah 11."

Mark was not the only Israelite to see this happening. Ellen Davis, *Proverbs, Ecclesiastes, Song of Songs* (Louisville, KY: Westminster John Knox, 2000), 231, observes: "Most briefly stated, the Song is about repairing the damage done by the first disobedience in Eden, what Christian tradition calls 'the Fall.'" N. T. Wright, *The New Testament and the People of God* (Minneapolis: Fortress, 1992), 262, also asserts that first-century CE Israelites saw the call of Abraham as the beginning of God's work of reversing the curse of evil initiated by Adam.

Christian Israelites continued this tradition. In his comment on Romans, 8:20–23, where Paul speaks of the redemption of creation and the human body from corruption, Robert Jewett in *Romans* (Minneapolis: Fortress, 2007), 515, says: "this Edenic hope features the converted 'children of God' . . . whose ministration would overturn the Adamic curse." See also Cynthia Long Westfa, *Paul and Gender* (Grand Rapids, MI: Baker Academic, 2016), 61: "He is deeply interested in . . . how Jesus reversed the effects of the fall."

4. Andrew Elphinstone, *Freedom, Suffering and Love* (London: SCM, 1976), 6; Judy Cannato, *Field of Compassion: How the New Cosmology Is Transforming Spiritual Life* (Notre Dame, IN: Sorin Books, 2010), 58, 71.

5. Denis Edwards, "Teilhard's Vision as Agenda for Rahner's Christology," in *From Teilhard to Omega: Co-Creating an Unfinished Universe*, ed. Ilia Delio (Maryknoll, NY: Orbis, 2014), 56.

6. Cannato, *Field of Compassion*, 25–40.

## 1. LIFTING THE CURSE

1. Scholars have convincingly distinguished two different creation stories in Genesis. The first account in Genesis 1 is a liturgical celebration of the entire cosmos. The second begins in Genesis 2:4 and narrows the focus to the human creatures. Our narrative here follows more closely the second creation story.

2. Delio, *Christ in Evolution* (Maryknoll, NY: Orbis, 2008), 50.

3. The medieval carol "Adam lay i-bowndyn" expresses the view that, had Adam not eaten the apple, we and the world would be the poorer. The lines "Ne hadde never our lady / a ben hevene quen" carry the implication that had there not been a fall from grace in the garden, there would have been no Christ. See Thomas Wright, ed., *Songs and Carols from a Manuscript in the British Museum of the Fifteenth Century* (London: T. Richards, 1856), 32–33.

4. Pamela Eisenbaum, *Paul Was Not a Christian* (New York: HarperOne, 2009), 73. In Jewish thinking, idolatry, which is the decision to turn away from the Creator to the

worship of created things, is the "primary cause of every other sin." Maximus the Confessor, *Ambiguum* 41, 6 in *On Difficulties in the Church Fathers: The Ambigua*, ed. and trans. Nicholas Constas, Vol. 2 (Cambridge, MA: Harvard University Press, 2014), says the same thing: "But moving naturally, as he was created to do, around the unmoved, as his own beginning (by which I mean God), was not what man did. Instead, contrary to nature, he willingly and foolishly moved around things below him, which God commanded him to have *dominion over.*"

5. David Vincent Meconi, in "Establishing an I-Thou Relationship between Creator and Creature," in *On Earth as It Is in Heaven*, ed. David Meconi (Grand Rapids, MI: William B. Eerdmans, 2016), 280, observes that God creates in the "third-person" manner but thereafter engages with creation in the "second person." God must command into being, "but he then refuses to domineer mere things but chooses, rather, to coax every creature personally."

6. The French Dominican priest and Biblical scholar Dominique Barthélemy, in *God and His Image* (San Francisco: Ignatius, 2004), 23–25, interprets the first couple's choice in a similar manner and offers a helpful qualification about the nature of their choice. Eating forbidden fruit is not their first decision to do something evil. Prior to that act, they decide to be the arbiters of what constitutes good and evil and to rely solely upon their own experience of what is distasteful or delightful. So there is no court of judgment other than their own experience. There is no "conscience" that calls into question the couple's own experience of pleasure or distaste. Barthélemy contrasts this with the wide-awake human conscience that remains lucid and alerts a person to having done wrong. Barthélemy concludes from this that the forbidden fruit cannot be the actual experience of evil as well as good. If humans are truly free, they must be free to experience evil or they cannot truly choose to do good. An individual's choice to do good, to be faithful to God, may legitimately come in the wake of being unfaithful if that choice proceeds from an awakened conscience. To clarify the meaning of the eating of the forbidden fruit, Barthélemy contrasts one who commits evil but retains a conscience with one who consents to the evil done by another. We might think that those who actually do evil are worse than those who consent to the evils that others may do. But such is not the case. Those who commit evil may do so in spite of themselves and upon reflection repent. But those who consent to the evil committed by others have stifled conscience, judging to be good that which is evil.

7. David Bentley Hart, *The Beauty of the Infinite* (Grand Rapids, MI: William B. Eerdmans, 2003), 20.

8. Hart, *The Beauty of the Infinite*, 143.

9. *Ambiguum* 7, 5.

10. Torstein T. Tollefsen, "Christocentric Cosmology," in *The Oxford Handbook of Maximus the Confessor*, eds. Pauline Allen and Bronwen Neil (Oxford: Oxford University, 2015), 319:

> From the way Maximus presents his cosmic view . . . one gets the impression that nothing is excluded, except corruptibility and sin. . . . Maximus . . . focuses on human beings. He never says explicitly that animals, vegetation, natural elements, and minerals have any place in the soteriological scheme. However, the way he contemplates the world in its divine roots (the *logoi*), . . . it seems in principle impossible that any being at all is wasted. . . . [T]he temporal order . . . is not some kind of theatrical show of just temporary value. This created whole . . . is not made to be annihilated, but rather to participate in the universal transfiguration and glorification.

11. Joseph Nguyen, in *Apatheia in the Christian Tradition* (Eugene, OR: Cascade Books, 2018), 11, describes the view of the influential fourth-century theologian Evagrius Ponticus, who adapted the views of Origen of Alexandria (185–253 CE). In Origen's view, God created souls before the material world. These souls forgot they were created in God's image and fell away into the created world. So God punished them by giving them bodies. Evagrius's adaptation of Origen's view led him to "perceive the human body as a mere means of the soul's spiritual journey back to God. For Evagrius, the ultimate aim of spiritual practice is the return of the soul to God in a complete detachment from the bodily senses and the material world."

12. Juergen Moltmann, *God in Creation* (Minneapolis: Fortress Press, 1993), 200–203.

13. Carlo Rovelli, *Seven Lessons on Physics* (New York: Riverhead, 2014), 32–33.

14. The Book of Wisdom is a first-century BCE work included in the Greek canon. Wisdom 1:12–2:24 alludes to Genesis 3 and explicitly says that death entered the world when humans allied themselves with the devil and invited death with their wickedness.

15. Richard Bauckham, *The Bible and Ecology* (Waco, TX: Baylor University Press: 2010), 23, observes that the original vision for the relationship between humans and the wild beasts in Genesis was one of harmony. Humans and all living creatures were to be vegetarians. Even after God gives humans dominion over all living creatures, no permission is given to eat them. Only plants and fruit are to be eaten (Genesis 1:28–30). There is a change after the flood in what God authorizes humans to eat. Violence on the part of both humans and animals is accepted as a fact of life but given limits. In Genesis 9:1–6 God gives humans permission to kill and eat creatures of land, sea, and air, but not without first draining the lifeblood. Genesis 9:5–6 prohibits humans from taking human life. Verse 5 also implies that animals will turn against humans and even be required to give

an account for killing humans. The reciprocal violence between humans and animals is already foreseen in the curse in Genesis 3:15.

16. God's choice of Noah was the first strategy in Genesis for restoring harmony to the earth. "This one shall bring us relief from our work and the toil of our hands, out of the very ground that the LORD has put under a curse" (Genesis 5:29). This first strategy failed; beginning in youth the desires of human hearts continued to be full of evil inclinations (Genesis 8:21). God came up with a new strategy. God called into being through Abraham and Sarah a new people, Israel, who would receive God's Torah, and be a holy people. When this people also failed, God did not destroy them. God restored them to holiness through the sacrificial cult and the warnings and exhortations of the prophets.

17. N. T. Wright, *The New Testament and the People of God* (Minneapolis: Fortress, 1992), 167, 268–272.

18. Isaac Watts (1674–1748) describes Jesus's mission as a lifting of the curse in one of the verses of his popular Christmas carol, "Joy to the World": "No more let sins and sorrows grow, nor thorns infest the ground; he comes to make his blessings flow far as the curse is found."

19. Cannato, *Field of Compassion*, 6–7.

20. Delio, *The Unbearable Wholeness of Being*, 52.

21. Cannato, *Field of Compassion*, 48.

22. Jesus is pathbreaking, not in the sense of doing something no one had ever done, but in the sense of taking up a number of impulses for which there were precedents in Israel's traditions and culture and incorporating them into his vision of the dawning reign of Israel's God.

23. See Michael Aquilina and James L. Papandrea, *Seven Revolutions: How Christianity Changed the World and Can Change It Again* (New York: Image, 2015).

24. For a detailed account of the connectedness of everything in the universe resulting from the evolution of everything out of the Big Bang, see Cannato, *Radical Amazement* (Notre Dame, IN: Sorin Books, 2006), 41–44. She quotes Bede Griffiths's summarizing conclusion:

> The explosion of matter in the universe fifteen billion years ago is present to all of us. Each one of us is part of the effect of that one original explosion such that, in our unconscious, we are linked up with the very beginning of the universe and with the matter of the universe from the earliest stages of its formation. In that sense the universe is within us.

This same understanding is expressed in Islamic tradition. See Seyyed Hossein Nasr et al., eds., *The Study Quran* (New York: HarperOne, 2015), 4:

> According to a famous saying attributed to 'Ali ibn Talib (d. 40/661), the Prophet's cousin and son-in-law, who became the first Imam of Shiite Islam (632–61) and the fourth Caliph of Sunni Islam (656–61), "The whole of the Quran is contained in the *Fatihah*, the whole of the *Fatihah* in the *basmalah* ['In the Name of God, the Compassionate, the Merciful'], the whole of the *basmalah* in the *ba'* [the opening letter], and the whole of the *ba'* in the diacritical point under the *ba'*." This point can be understood to represent the first drop of ink from the Divine Pen (*al-qalam*; see 68:1c; 96:4c) with which God wrote the archetypes of all things upon the *Preserved Tablet* (*al-lawh al-mahfuz*; see 85:22c) before their descent into the realm of creation. In this sense, just as the *basmalah* marks the beginning of the Quran, so too does it mark the beginning of creation.

The Arabic letter *ba'* looks like this: ب . Note the position of the dot.

25. For a complete account of Lorenz's experience, see James Gleick, *Chaos* (New York: Penguin, 1987), 9–31.

## 2. THE GOSPEL OF MARK

1. Michael Lerner, *Jewish Renewal* (New York: HarperPerennial, 1994), 26:

> Throughout much of recorded history the oppressed have been socialized to believe that cruelty and oppression are inevitable, an ontological necessity, part of the structure of reality. Spirituality for them became identified with reconciliation to a world of oppression: either through learning to "flow" with the world as it is or through imagining that the material world in which they lived was a prelude to some higher nonmaterial world, and that the task of the living was to escape material reality into this spiritual realm which embodied the purity and deeper reality that could not be imagined on this earth.
>
> What the Jews heard was a very different message: that this world could be fundamentally transformed. . . . "[V]ery hard" is different from "impossible." . . . [It] might take thousands of years. But from the standpoint of this Jewish sensibility, what we do, how we live, the kind of society we build, can contribute to the defeat of cruelty.

2. Juergen Moltmann, *The Way of Jesus Christ* (New York: HarperCollins, 1990), 274:

> [T]he cosmic christology of the Epistles to the Ephesians and Colossians was dismissed by modern European theology as mythology and speculation. Anthropological christology fitted the modern paradigm "history," and itself unintentionally became one factor in the modern destruction of nature; for the modern reduction of salvation to the salvation of the soul, or to authentic human existence, unconsciously abandoned nature to its disastrous exploitation by human beings.

3. Within orthodox circles this reinterpretation shows up in *Te Deum Laudamus*, an early fourth-century CE hymn. N. T. Wright, *How God Became King* (New York: HarperOne, 2012), 43, comments:

> There we find the clause (in the translation of the Book of Common Prayer): "When thou hadst overcome the sharpness of death, thou didst open the kingdom of heaven to all believers." Read Matthew's gospel with *that* in mind, and you are almost bound to see the "kingdom of heaven" as a place from which believers might have been barred because of sin, but to which now, through the death of Jesus, they have access. What's more, though the hymn does not exactly say so, it hints at a parallel: Jesus opened the "kingdom" through his death, so it is presumably through and after death that believers enter this "kingdom" themselves.

Gerhard Lohfink, *Is This All There Is? On Resurrection and Eternal Life* (Collegeville, MN: Liturgical Press Academic, 2017), 12–13, quotes a third-century CE tomb inscription that expresses the belief that the soul of a young man has left the body and cares of a bitter life to "hurry on the divine road so that she [the soul, *psyche*] might arrive purified." The anchor carved beneath the inscription shows that it was for a Christian. Lohfink remarks that, while there were differences between the pagan and the Christian view, aspects of the view that the body was a tomb from which the soul found release in death "entered into Christianity and burrowed into many corners and crannies, despite the fact that it is incompatible with the Christian idea of creation and redemption."

4. Paul M. Blowers, "Unfinished Creative Business," in Meconi, ed., *On Earth as It Is in Heaven*, 188, describes Maximus's view:

> The "contemplation of nature," as he calls it, is less about reaching a transcendent vision beyond corporeality than preparing oneself through knowledge and virtue for the eschatological transformation of creation.... As for the practical dimension of human stewardship in creation, Maximus' signature contribution is his emphasis on the mediating work of humanity analogous to Christ's own work as cosmic mediator and reconciler.... [H]uman beings are uniquely ontologically positioned to cooperate with Christ in the healing and perfecting of creation, a work still in progress.

A most impressive description of this unifying and reconciling work is expressed in *Ambiguum* 41, 2–5. Delio, *Christ in Evolution*, 47–65, summarizes this history.

5. N. T. Wright, "Narrative Theology," in Katharine J. Dell and Paul M. Joyce, eds., *Biblical Interpretation and Method: Essays in Honour of John Barton* (Oxford: Oxford University Press, 2013), 190–192, adds to the precedent of the successful Maccabean Revolt the widespread popularity of the Book of Daniel, which provided a chronology for the arrival of the redemption promised in Isaiah, Zechariah, and Malachi. From Daniel, in

the period between the reign of Herod the Great and the outbreak of the Great Revolt in 66 BCE, many Israelites concluded the time for redemption had come.

6. Thomas E Boomershine, *The Messiah of Peace: A Performance-Criticism Commentary on Mark's Passion-Resurrection Narrative* (Eugene, OR: Cascade Books, 2015).

7. A new picture is evolving of how Christianity and Judaism emerged as two distinct religious communities. Across the Roman Empire, Israelites were relating their ideas and practices to those of their Gentile neighbors. Two scholars are helpful here. Boyarin, in *The Ways That Never Parted*, 74, suggests the picture not of a family tree that grows only by increasing divergence, but rather the picture of converging ripples caused by pebbles thrown into a pond. Some of these pebbles were the various early Christian communities; others were non-Christians like Philo. Mark's story was one of those pebbles.

Rodney Stark, *The Rise of Christianity: How the Obscure, Marginal Jesus Movement Became the Dominant Religious Force in the Western World in a Few Centuries* (San Francisco: HarperCollins, 1997), 49, 57–59, explains how the Christian solution to the relationship between Israelite and Gentile culture appealed to many Israelites:

> [N]ot only was it the Jews of the diaspora who provided the initial basis for church growth during the first and early second centuries, but . . . Jews continued as a significant source of Christian converts until at least as late as the fourth century and . . . Jewish Christianity was still significant in the fifth century.

Stark attributes this to the fact that diaspora Jews "were in the unstable and uncomfortable condition of social marginality."

8. Boomershine, *The Messiah of Peace,* 134–135. Joel Marcus, *Mark 8–16* (New Haven, CT: Yale University Press, 2009), 783–784, observes that in addition to the Greek word *lestes* (Mark 11:17; 15:27), the word *skeuos* (Mark 11:16) would also have reminded Mark's audience of the Great Revolt taking place in their present or immediately past experience. It is translated by the New American Bible Revised Edition (NABRE) as "anything" in the clause "did not permit anyone to carry anything through the temple area"; as "any merchandise" by the New English Translation (NET) Bible; and as "any vessel" by the King James Version (KJV). But it can also be a "weapon" (Genesis 27:3, "now take your *skeuos*, your arrow quiver and your bow"; Deuteronomy 1:41, "your military *skeue*"). Along with the word *lestes* it "would probably remind the Markan community of the recent intrusion of armed revolutionaries into the Temple precincts."

9. That Mark preached in Rome and Egypt would not mean that he preached to Gentiles. Both Rome and Alexandria had significant Jewish populations that included Christians. See Stark, *The Rise of Christianity*, 57, 68, and Peter Lampe, *From Paul to Valentinus* (Minneapolis: Fortress, 2003), 38–40.

10. This translation is from Philip Schaff and Henry Wace, eds., *A Select Library of Nicene and Post-Nicene Fathers of the Christian Church,* Series 2, Vol. 1 (New York: Christian Literature Company, 1890), 172–3. All the bracketed identifications except the last are from Richard Bauckham, *Jesus and the Eyewitnesses* (Grand Rapids, MI: Eerdmans, 2006), 203.

11. Schaff and Wace, eds., *A Select Library*, 116.

12. For a review of the contrasting stances towards the historical reliability of the Jesus traditions in the New Testament and the work of the Jesus Seminar in particular, see N. T. Wright, *Jesus and the Victory of God* (Minneapolis: Fortress, 1996), 21, 29–35.

13. Other instances are Mark 3:20–35; 5:21–43; 6:6b–33; and 14:1–11. In addition, Mark 15:53–72 starts with leading Jesus to the high priest, switches to Peter's following, then cuts to the trial before the high priest, then back to Peter's denials.

14. K. L. Schmidt, *Die Rahmen der Geschichte Jesu* (Berlin: Trowitzsch & Sohn, 1919), 281. Schmidt is often quoted as having said a "string of pearls," as though Schmidt were picturing an orderly succession of stories and teachings. See the reference to Schmidt by James D. G. Dunn, "Form Criticism," in Paula Gooder, ed., *Searching for Meaning: An Introduction to Interpreting the New Testament* (Louisville, KY: Westminster John Knox, 2009), 23. Sharyn Dowd and Elizabeth Struthers Malbon also employ the phrase in "The Significance of Jesus' Death in Mark: Narrative Context and Authorial Audience," *Journal of Biblical Literature* 125, 2 (2006): 278. In fact, Schmidt explicitly rejects that image in favor of another image. He says "the Markan presentation is not a string of pearls loosely lined up one after the other [*nicht eine Perlenkette von lose aneinandergereihten Perlen*], between which one can insert others here and there, but a heap of disordered pearls [*ein Haufe von nicht aufgereihten Perlen*), some of which are connected to each other." Dowd and Malbon describe the way many of Mark's "pearls" are connected to each other.

15. Erich Auerbach, *Mimesis: The Representation of Reality in Western Literature* (Princeton, NJ: Princeton University Press, 1953), 12.

16. See Raymond E. Brown, *An Introduction to the New Testament* (New York: Doubleday, 1997), 371–373, for an account of the changes in how scholars have viewed the cultural background of the Gospel according to John. The critical sentence is: "A very different approach would see the basic origins of Johannine Christianity within that Palestinian world with all its Jewish diversity—a world that had been influenced by Hellenism but where reflection on the heritage of Israel was the primary catalyst."

17. N. T. Wright, *Surprised by Hope* (New York: HarperOne, 2008) provides a full description of how the hope for the future in the Bible is the restoration of creation to health. He frequently draws on Paul's letters, showing them to be a source of this hope.

See also chapter 15, "The Reign of God," in Marjorie Hewitt Suchocki's *God Christ Church* (New York: Crossroad, 1992).

18. Barbara R. Rossing, *The Rapture Exposed* (New York: Basic Books, 2004), provides an excellent introduction to interpreting Revelation in the context of Israel's traditions and the culture of the Roman Empire. She also explains the problems with the dispensationalist interpretation that is so popular in the United States today.

## 3. "HE WAS AMONG THE WILD BEASTS"

1. Luke begins and ends his gospel in Jerusalem. He also begins the mission of Jesus's disciples in Acts in Jerusalem and it radiates out from there. The final temptation in Luke takes place in Jerusalem. Matthew's gospel ends with Jesus telling his disciples, "All power in heaven and on earth has been given to me. Go, therefore, and make disciples of all nations" (Matthew 28:18–19). In this way, the devil's temptation to worship him in order to gain all the kingdoms of the world is ironically fulfilled when Jesus remains faithful to God rather than worshipping the devil.

2. Mark's strategy of telling the listeners just enough to provoke the imagination fits that commended by the late first-century BCE grammarian Demetrius. See N.G.L. Hammond and H. H. Scullard, eds., *The Oxford Classical Dictionary* (Oxford: Clarendon, 1970), 326. Demetrius relies on an earlier authority, Theophrastus, quoted in Margaret Ellen Lee and Bernard Brandon Scott, *Sound Mapping the New Testament* (Salem, OR: Polebridge Press, 2009), 80:

> You should not elaborate everything in punctilious detail but should omit some points for the listener to infer and work out for himself. For when he infers what you have omitted, he is not just listening to you but he becomes your witness and reacts more favorably to you. For he is made aware of his own intelligence through you, who have given him the opportunity to be intelligent. To tell your listener every detail as though he were a fool seems to judge him one. (*Elocutione 222*)

Mark does not leave his listeners without guidance for how to work things out. He saturates his narrative with allusions to traditions and experiences his Israelite listeners are equipped to recognize. As he goes along, he repeats words and phrases that gather associations as they weave through the story.

3. Mark's Greek term for "wild beasts," *therion*, matches the Septuagint translation of Genesis 2:19.

4. In Bauckham's exposition of Mark's temptation story, he makes an explicit connection between Isaiah and Genesis: "the return to Eden we find in Isaiah 11." See Bauckham, *The Bible and Ecology*, 127.

5. Two popular first-century BCE compositions set a precedent for Mark's substitution of Satan for the snake. Wisdom 2:24 says, "But by the envy of the devil, death entered the world." And 1 Enoch 69:6 makes Gadre'el, a fallen angel, "the one who showed all the blows of death to the sons of men, and . . . led Eve astray." Translation from George W. E. Nicklesburg and James C. VanderKam, *1 Enoch 2: A Commentary on the Book of Enoch* (Minneapolis: Fortress, 2012), 297. The Book of Revelation, composed by another Christian who was also an Israelite, speaks of "the ancient serpent, who is called the Devil and Satan, who deceived the whole world" (12:9), suggesting that Israelites listening to Mark might easily have made the connection.

6. Scholars widely agree that this second passage is by a sixth-century BCE successor of the eighth-century BCE prophet who composed 11:6–9. Isaiah 65:25 excerpts the earlier vision, but imagines the "Peaceable Kingdom" as a manifestation of God's work of "creating new heavens/ and a new earth" (65:17). Isaiah 65:25 says:

> The wolf and the lamb shall pasture together,
> and the lion shall eat hay like the ox—
> but the serpent's food shall be dust.
> None shall harm or destroy
> on all my holy mountain, says the LORD.

It is intriguing that in the transformed heaven and earth envisioned in Isaiah 65:17–25 this later Isaiah does not see God lifting the curse pronounced upon the serpent in Genesis 3:14–15. The serpent will still eat dust.

7. David Bentley Hart, *The Beauty of the Infinite* (Grand Rapids, MI: William B. Eerdmans, 2003), describes the history of Western philosophy as a history of being as *necessity*, in contrast to the Christian theological tradition flowing out of Israel's traditions and founded on a recognition of the utter *gratuity* of being. William Desmond, *God and the Between* (Oxford: Blackwell, 2008), 27, echoes Hart's assessment in his remark that Nietzsche sought "to overcome the discordance of being and good by affirming all being as 'good'" simply "as it is." Desmond raises the question, "[D]oes 'It is so' now become 'It must be so'?" For Israel what is now does *not have to be*. Consequently, something quite other than what is now may have been and may be again. Incredibly the Israelites could imagine that the human practice of killing and eating animals and the practice of animals eating other animals, though indubitably the way things are now, does not have

to be the way things will be or were at one time in the past. The way things are now is not by necessity but only by happenstance and therefore can change. Juergen Moltmann notes the connection between Mark's picture of Jesus with the wild beasts in 1:13 and Isaiah's vision in 11:6–8. He affirms Israel's capacity for imagining the present conditions prevailing on earth as unnecessary, yielding to a transformed order in which there will be "peace for humankind and animals, where there is no killing anymore. Children will play with snakes," lifting the curse of Genesis 3:15 where the relationship between humans and snakes will be one of violence, "and lions will eat straw (Isaiah 11:6–8). This indicates that meat-eating humans and the beasts of prey are peaceless creatures. The future doesn't belong to them. The messianic kingdom is without violence. The present situation in the human and animal world is not in the order of their creator, though some say fight and competition with the result of the survival of the fittest is the given law of nature." Moltmann, "Epilogue," in *Turning to the Heavens and the Earth: Theological Reflections on a Cosmological Conversion*, Julia Brumbaugh and Natalia Imperatori-Lee, eds. (Collegeville, MN: Liturgical Press, 2016), 257.

8. Richard Bauckham, *The Bible and Ecology*, 127, sees the wild beasts as ambiguous characters poised between a clear enemy, Satan, and clear friends, the angels. These are "enemies of whom Jesus makes friends." "Satan is the natural enemy of the righteous person and can only be resisted. Angels are the natural friends of the righteous person: they minister to Jesus. But between Satan and the angels the wild beasts are more ambiguous. On the basis of the common perception of wild beasts as a threat to humans, we might expect them to be dangerous enemies, especially when located in the wilderness, the habitat that belongs to them and not to humans. But, on the other hand, since Jesus is the messianic king, inaugurating his Kingdom, might we not expect his relationship to the wild animals to be appropriate to that Kingdom, the return to Eden we find in Isaiah 11?" Bauckham points out that Mark's phrase "with the wild beasts" (NABRE translates *meta* as "among" in Mark 1:13) involves no suggestion of hostility. Whenever Mark speaks of a person as being "with" (*meta*) someone else, it is "in the sense of close, friendly association." (See Mark 3:14; 5:18; 14:67, where the NABRE translates *meta* as "with.") The same is true of descriptions of Noah's relationship with the various animals where the Greek translation uses the same phrase (Genesis 7:23; 8:1, 17; 9:12).

The bracketing of the beasts with Satan and the angels generates a second level of meaning to the beasts in this brief story. Are these beasts internal temptations as well as Adam's companions before the creation of Eve and before the Fall? The listeners are provoked to wonder by the presence of the tempter. When Satan reappears in the form of Simon Peter in the middle of the story, signs of Jesus's internal struggle come to the fore. (See Mark 8:31–33.)

9. To imagine what choices Jesus must have wrestled with during his forty days in the wilderness is not to be guilty of psychologizing. The sparseness of detail in Mark's account provokes us to thought time and again. Mark's narrative is, in the phrase of Erich Auerbach, "fraught with background." We search out that background on the basis of what Mark tells us in the story. One thing that becomes clear as the story moves along is that Jesus is not married and does not marry. This was not unprecedented for an Israelite, but it was unusual. Paul was not married (1 Corinthians 9:5). Observing this usual state, the listeners may well imagine that he had to work through the question whether he should marry or not. What better time to have done it than during his forty days in the wilderness? What other temptations does the narrative imply he had to have worked through? Erich Auerbach, *Mimesis* (Princeton, NJ: Princeton University Press, 1953), 12.

10. Eleonore Stump, *Wandering in Darkness: Narrative and the Problem of Suffering* (Oxford: Clarendon, 2010), 126.

11. Gerhard Lohfink, *Jesus of Nazareth* (Collegeville, MN: Liturgical Press, 2012), 93.

12. John J. Collins, *The Scepter and the Star: The Messiahs of the Dead Sea Scrolls and Other Ancient Literature* (New York: Doubleday, 1995), 68: "This concept of the Davidic messiah as the *warrior king* who would destroy the enemies of Israel and institute an era of unending peace constitutes the *common core of Jewish messianism around the turn of the era*." [Italics added.]

13. Bauckham, *The Bible and Ecology*, 128, notes that the human threat to animals today can be seen when we observe how many wild beasts Jesus may have encountered in the wilderness have become extinct in Palestine over the past one hundred years: the wild ass, the desert oryx, the addas, the ostrich. Bauckham also points out that in the story of Mark the phrase "with the wild beasts" means to be in a friendly relationship with them. One might think that since Jesus is a messianic king, this relationship with animals would be one of domination in order to make them useful for humans. But the use of the phrase "to be with someone" elsewhere in Mark and in the Greek translation of the story of Noah and the animals does not allow such a meaning.

14. Barbara R. Rossing, *The Rapture Exposed: The Message of Hope in the Book of Revelation* (New York: Basic Books, 2004), 7, notes that conservative commentator Ann Coulter puts in words the attitude Western industrial civilization has so often exhibited in action. Coulter argues that this predatory behavior is mandated by the God of the Bible: "God gave us the earth. We have domination over the plants, the animals, the trees. God said, 'Earth is yours. Take it. Rape it. It's yours.'"

15. Bauckham, *The Bible and Ecology*, 128–129, observes, "Jesus does not adopt the animals into the human world, but lets them be themselves in peace, *leaving them in the*

*wilderness*, affirming them as creatures who share the world with us in the community of God's creation" (italics added). He describes Jesus's relationship with the animals as a "biblical symbol of the human possibility of living fraternally with other living creatures. Like all aspects of Jesus' inauguration of the Kingdom of God, its fullness will be realized only in the eschatological future, but it can be significantly anticipated in the present by respecting wild animals and preserving their habitat."

16. The healing of the natural environment is not the central focus of Jesus's ministry in comparison with the healing of human bodies and spirits and society's divisions. But it is essential to the lifting of the curse.

17. Martin Luther, *Commentary on Romans*, trans. J. Theodore Mueller (Grand Rapids, MI: Zondervan, 1954), xvii: "Faith is a divine work in us. It changes us and makes us to be born anew of God (John 1). It kills the old Adam and makes altogether different people, in heart and spirit and mind and powers, and it brings with it the Holy Spirit. Oh, it is a living, busy, active, mighty thing, this faith. And so it is impossible for it not to do good works incessantly. It does not ask whether there are good works to do, but before the question rises, it has already done them, and is always at the doing of them."

18. Rossing's *The Rapture Exposed* provides a very readable scholarly assessment of dispensational theology together with an account of how its proponents have marketed it. James M. Efird, *End-Times* (Nashville: Abingdon, 1986), 22–24, describes the political, economic, and scientific upheaval and disenchantment with the established church in the late seventeenth and early eighteenth centuries, as well as the personal experience that led Darby to develop his innovative interpretation of the Bible.

19. Juergen Moltmann, *The Coming of God: Christian Eschatology*, trans. Margaret Kohl (Minneapolis: Fortress, 1996), 268–270, notes that the leading Christian theologians, beginning with Irenaeus of Lyons in the second century and stretching from Augustine in the fourth century, Gregory the Great and Maximus the Confessor in the seventh century, to Aquinas in the thirteenth century, through all medieval theology, right down to present-day Roman Catholic dogmatics, all affirmed the permanent value of the material creation, anticipating its transformation, not its annihilation. The Lutheran orthodoxy of the seventeenth century was, in contrast, unanimous in maintaining that the ultimate fate of the material world was annihilation. All that would remain would be heaven and hell, the abodes of the saved and the damned.

20. Hal Lindsey, *The Late Great Planet Earth* (Grand Rapids, MI: Zondervan, 1970).

21. N. T. Wright, *Jesus and the Victory of God* (Minneapolis: Fortress, 1996), 513.

22. In Jesus's three prophecies of the coming of the Son of Man at some future time (Mark 8:38; 13:26–27; 14:62), he does not explicitly say that the Son of Man will receive dominion over the nations. If listeners detect the allusion to Daniel 7 in the image of the

Son of Man coming in clouds, will they remember that in Daniel he receives dominion over the nations? N. T. Wright, *How God Became King*, 139, 192, speaks of how the term "son of man" would evoke the whole narrative for a scripture-soaked audience; he also asserts Daniel was a popular text for Israelites longing for the nation's redemption. Mark's story of the reconciliation of Israel with the Gentiles could encourage Israelite listeners to remember that part of Daniel's vision. Mark 13:27 is particularly suggestive: "the Son of Man will send out his angels and gather his elect from the four winds." Wright observes in *Jesus and the Victory of God*, 363, that the fall of Jerusalem would vindicate the Son of Man and "be the sign that the followers of this 'son of man' would now spread throughout the world: his 'angels,' that is, messengers, would summon people from all compass points to sit down with Abraham, Isaac and Jacob in the kingdom of YHWH."

23. Gerhard Lohfink, *Jesus of Nazareth: What He Wanted, Who He Was* (Collegeville, MN: Liturgical Press, 2012), 43; also Moltmann, *The Coming of God*, 94.

24. Mark, or Jesus himself, achieves a particular interpretation of how the Son of Man receives dominion over the nations by creating the backstory that leads to the coming of the Son of Man in the future. The backstory has two stages. First, Jesus is the Son of Man who brings healing to creation. (Mark 2:10; 28. Matthew, Luke, and the Q tradition contain additional sayings of Jesus that belong in this first stage.) Second, this Son of Man will suffer, die, and rise from the dead (principally Mark 8:31; 9:31; 10:33–34). It is clear from the narrative that in Jesus the Son of Man has rejected violence as the path to ruling the nations. When in the third stage the Son of Man comes with the clouds in power and glory (principally Mark 8:38; 13:26–27; 14:62), it will be to rule humanely after an earthly ministry of healing every manifestation of violence introduced into the world by Adam and Eve's rejection of God's rule.

25. Delio, *Christ in Evolution*.

## 4. JESUS AND WOMEN

1. David Rhoads, Johanna Dewey, and Donald Michie, *Mark as Story: An Introduction to the Narrative of a Gospel* (Minneapolis: Fortress, 2012), xi: "It has become clear that Mark's story was presented from memory, told all at one time, probably in houses, in marketplaces, at meals, at evening gatherings, and at synagogue-like assemblies."

2. Exceptions to Mark's otherwise minimalist style are Mark 2:1–12; 6:17–29; 9:14–29. The graphic descriptions are striking: tearing up the roof and lowering the paralyzed man through the hole, Herod's banquet and his alcohol- and hormone-induced promise to Herodias's daughter, and repeated descriptions of the violence done by the spirit to the father's son. In his versions of these stories, Matthew has eliminated all or much of

Mark's detail. Luke kept but modified the description of lowering the paralyzed man in the first story, probably in line with the different building practices of his listeners (Luke 5:19). He does not tell the story of Herod's banquet at all. He shortens the story of the healing of the boy with the violent spirit very much as Matthew does. He eliminates the exchange between Jesus and the father initiated by Jesus's question, "How long has he been like this?" He also omits the report that after Jesus cast out the spirit the boy appeared to be dead and Jesus had to raise him up (Luke 9:37–43; Matthew 17:14–20). In Mark's story, the boy's apparent death and resurrection foreshadows Jesus's own death and resurrection. Mark uses the verb *egeiro* for Jesus raising the boy up and *anistemi* for the boy's standing up (9:27). The young man at Jesus's tomb again uses *egeiro* to declare to the women that Jesus "has been raised" (16:6). Jesus employs *anistemi* in the three passion prophecies (8:31, 9:31, and 10:34) to speak of his resurrection. The way these verbs foreshadow Jesus's resurrection no doubt explains why this is the only healing miracle in the second half of Mark apart from the two stories of healing blind men that bracket the journey narrative.

3. Sharon Dowd and Elizabeth Struthers Malbon, "The Significance of Jesus' Death in Mark: Narrative Context and Authorial Audience," in *Journal of Biblical Literature* 125, 2 (2006): 278, describe the careful structuring in Mark 8:22–10:52: "The development of this section clearly involves more than the stringing of pearls. Each passion prediction (8:31; 9:31; 10:33–34) has been elaborated into a passion prediction unit by the addition of misunderstanding by the disciples and Jesus' discipleship instruction in response." These exhibit the narrator's acute sense of artistic ordering of his received traditions. This is not necessarily the result of textual editing. It could just as well be evidence of an oral narrator who has retold his story repeatedly and has developed a pattern retainable in his memory. An oral narrator who has structured or learned this repeating pattern from another storyteller would be careful to perform it in such a way that the audience would feel the impact of returning to the themes once more and enjoy the variation. As Dowd and Malbon say, 272–273, "Mark's Gospel is a story, a sequential narration of events with beginning, middle, and end designed to be *heard* [italics added] in that order. . . . In addition, Mark's story is designed not for the eye but for the ear. Because the first-century context of this whole Gospel was oral, we will listen for echoes (to use an oral metaphor) of earlier scenes."

4. Richard A Horsley, *Hearing the Whole Story: The Politics of Plot in Mark's Gospel* (Louisville, KY: Westminster John Knox, 2001), 203–229. For corroboration of Horsley's view, see Ulrich Luz, *Matthew 1–7: A Commentary* (Minneapolis: Augsburg, 2007), 246; Allen Verhey, *Remembering Jesus: Christian Community, Scripture, and the Moral Life* (Grand Rapids, MI: Eerdmans, 2002), 171–177; Philo, *The Special Laws* XXXI:

> Market-places and council halls and lawcourts and gatherings and meetings where a large number of people are assembled, and open-air life with full scope for discussion and action—all these are suitable to men [*andrasin*, the word for males in contrast to females] both in war and peace. The women are best suited to the indoor life which never strays from the house, within which the middle door is taken by the maidens [*parthenois*] as their boundary, and the outer door by those who have reached full womanhood [*teleiais gynaiksi*].

*Philo in Ten Volumes (and Two Supplementary Volumes)*, translated by F. H. Colson, Vol. VII (Cambridge, MA: Harvard University Press, 1937), 169.

5. We must beware of creating linguistic binaries that present one group as good and the other as somehow evil. Christianity has a long history of contrasting itself to Judaism accompanied by an entire lexicon of derogatory terms for describing Judaism.

6. A ninth episode, Mark 3:31–35, includes a woman, Jesus's mother. But she is in the company of Jesus's family as a whole. While as a family they come to seize him because they think he is out of his mind (Mark 3:19b–21), there is no direct interaction between Jesus and his mother. Nevertheless, the episode may be relevant to the role of women in the Christian movement. Horsley, *Hearing the Whole Story*, 224, points out that when Jesus declares, "whoever does the will of God is my brother and sister and mother," it is notable that there is no father in the list, on the one hand, and that "sister" is explicitly included, even though "brother" in Greek is generic and could include sisters. The same is true in Mark 10:30 where Jesus again lists brothers *and sisters* and mothers and children, but not fathers.

7. Despite the telling of the story of Jairus's daughter in conformity with patriarchal norms, Horsley, *Hearing the Whole Story*, 211–212, sees the detail of her age, twelve years old, along with the fact that the woman with the hemorrhage suffered for twelve years, as representing the renewal of Israel:

> They represent other women whose life circumstances and illness are similar. They are typical of peasant women involved in the Jesus movement and addressed in the Gospel. Insofar as they represent other women in similar circumstances, moreover, they represent the whole society, Israel. Just as the Twelve are appointed as representatives of Israel undergoing renewal and twelve baskets of leftover fragments are gathered from the great feeding of Israel in the wilderness . . . the twelve-year-old woman now near death, whose father is "one of the leaders of the synagogue" (local assembly of Israelites), and the other woman's twelve years of bleeding represent (the twelve tribes of) Israel in desperate condition. And their respective healings by aggressive action out of faith and restoration to life by Jesus represent the new life of Israel.

8. Joel Marcus, *Mark 1–8: A New Translation with Introduction and Commentary* (New Haven, CT: Yale University Press, 2000), 357.

9. Malachi 2:13–16 also asserts that God hates divorce, raising in v 15 the same view of marriage we hear in Genesis 2:24: "Did he not make them one, with flesh and spirit?"

Pamela Eisenbaum, *Paul Was Not a Christian* (New York: HarperOne, 2009), 121–122, points out that the community responsible for the Dead Sea Scrolls objected to divorce as Jesus did and on the basis of Genesis 1:27. The Covenant of Damascus 4:19–20 reads:

> The builders of the wall . . . have been caught by lust in two things:
> by marrying two women during their lifetime,
> whereas nature's principle is
> *Male and female created He them.*

Transl. G. Vermes in A. Dupont-Sommer, *The Essene Writings from Qumran* (Cleveland: Meridian, 1961), 128–129.

Eisenbaum offers reasons for understanding the "builders of the wall" to be Pharisees. Both the Covenant of Damascus and Jesus therefore criticize the Pharisees for permitting divorce and base their criticism on Genesis 1–2. Joel Marcus, *Mark 8–16* (New Haven, CT: Yale University Press, 2009), 700, reports that another Qumran text recognized divorce and that the Covenant of Damascus 4:19–20 may be prohibiting bigamy and perhaps also remarriage by a divorced man while his former wife is still alive. Regardless of its specific reference, it remains significant that the Covenant of Damascus refers to Genesis for its authority as Malachi and Jesus both do.

10. The possibility, to which Jesus objects in Mark 10:12, that a wife could divorce her husband reflects the Hellenistic context outside Palestine where Mark was probably telling his story. See 1 Corinthians 7:10–11; John H. Elliott, "Jesus Was Not an Egalitarian: A Critique of an Anachronistic and Idealist Theory," *Biblical Theology Bulletin* 32 (2002): 80. Ulrich Luz, *Matthew 1–7: A Commentary* (Minneapolis: Augsburg, 2007), 251–252, also comments that in Israel the law of Moses makes divorce a prerogative only for the male. In a footnote he cites a source for rare instances of a Jewish woman divorcing. Such women were normally from the upper classes.

11. See Marcus, *Mark 8–16*, 712: "To many, the Pauline position, with its exaltation of familial and communal harmony over purity considerations and its recognition that in some cases peace is best served by divorce, may seem more attractive than strict adherence to the Markan Jesus' verdict, which may appear to be based on an excessive anxiety about sexual defilement." We would disagree with Marcus that a concern about sexual defilement lies at the root of Jesus's rejection of divorce in Mark. Jesus's phrase "one flesh" to describe the couple goes beyond physical sex. It once again evokes the entire relationship they enjoyed before the curse. This relationship is not only sexual. Since the

curse introduced male domination, the original relationship between man and woman was free of domination. Divorce in Jesus's day is itself a manifestation of domination. The husband could divorce his wife without her consent. If he did so, she was without means of support in a patriarchal society where most women depended upon living in a family where the male was the means of support.

12. Marcus, *Mark 8–16*, 712–713, quotes David Adams's observation that "already in earliest Christianity the church saw fit to modify even dominical teaching on important subjects, a responsible freedom that, as precedent, is probably more significant than any one of the teachings as such." Adams draws attention to 1 Corinthians 9:14 where Paul reports a clear command of Jesus "that those who preach the gospel should live by the gospel." In the next sentence Paul declares for himself that he views Jesus's command not as a command but as an option: "I have not used any of these rights" (1 Corinthians 9:15). See also Nicholas Peter Harvey, *Morals and the Meaning of Jesus: Reflections on the Hard Sayings* (Cleveland: Pilgrim Press, 1993), 26–27: the aim of Jesus's mission was not to articulate standards and ideals but to stir, even convert the imagination, so that followers could identify with him "in the Spirit which made possible *the free play of their particular gifts*" in response to the many ambiguities that life would throw at them.

13. Elizabeth Struthers Malbon, *Mark's Jesus: Characterization as Narrative Christology* (Waco, TX: Baylor University, 2009), 225–226.

14. William Desmond, *The Intimate Universal: The Hidden Porosity among Religion, Art, Philosophy, and Politics* (New York: Columbia, 2016), 188: "agapeic service" is beyond both servility and "erotic sovereignty."

15. See Mark 1:31, where the NABRE translators see in the context that *diekonei* implies that Peter's mother-in-law "waited on them," that is, performed the service for men expected of women. Richard Rohr, *Simplicity: The Freedom of Letting Go* (New York: Crossroad, 2003), 26, asks, Why did Jesus not come as a woman? His answer is that "if Jesus had come as a woman and had this woman been forgiving and compassionate, and had she taught non-violence, we wouldn't have experienced that as revelation. 'Oh, well, a typical woman,' we would have said. But the fact that a man in a patriarchal society took on these qualities that we call 'feminine' was a breakthrough in revelation."

16. After Jesus healed her, Simon's mother-in-law served (*diekonei*) Jesus and his male disciples (Mark 1:31). Later in Mark's narrative, Jesus tells his male disciples that they must learn to serve as well, because he "did not come to be served but to serve" (Mark 10:45). Although they are male, they must learn to take on this traditionally female role. On first hearing Mark's story, a listener would probably not see anything more in the service Peter's mother-in-law renders than a woman carrying out the duty expected of her. But on a second hearing an attentive listener might hear in her service a prophetic fulfillment of the service Jesus will commend to his male disciples as well. By

commending *diakonia* to the male disciples, who have been sent out by Jesus to preach and drive out demons and heal the sick as he has been doing himself (Mark 6:12–13), it becomes clear that the service Jesus has in mind is not the traditional service rendered by women to men. Mark 15:40–41 reports that the women who followed Jesus in Galilee "served" him; their service may have included providing for the physical needs of Jesus and his male disciples out of their own resources. But Jesus's use of the word to describe the proper aspirations of the male disciples challenges listeners to expand their view of what it means to serve. They may take their cue from Jesus. It begins in bringing good news, healing, feeding, and reconciling. In comes to fullness in giving one's life for many.

17. A good set of cross-references to the other three gospels, like the set in the NABRE, will guide the reader to the passages in those gospels that probably influenced the person or persons who composed this Longer Ending. Larry Hurtado, *Lord Jesus Christ: Devotion to Jesus in Earliest Christianity* (Grand Rapids, MI: William B. Eerdmans, 2003), 585, observes that the Longer Ending is constructed from the other three gospels and only those three. This suggests that the collection of the four canonical gospels was completed and in circulation at the time this ending was composed. It was the very existence of the collection that highlighted what was missing in Mark. This new ending was composed by the second century, since the apologist Justin Martyr knew it.

18. Thomas E. Boomershine, "Mark 16.8 and the Apostolic Commission," in *Journal of Biblical Literature* 100 (1981).

19. See Ivone Gebara, *Out of the Depths: Women's Experience of Evil and Salvation* (Minneapolis: Fortress, 2002), 85–90, for a description and assessment of the different ways in which the Church has applied Jesus's modeling and teaching of sacrifice to women and to men to maintain the subordination of women to men.

20. Meyers, Carol. "Women in the OT," in *The New Interpreter's Dictionary of the Bible S–Z*, Vol. 5, ed. Katharine Doob Sakenfeld et al. (Nashville: Abingdon, 2009), 888–892.

21. Amy L. Wordelman, "Everyday Life: Women in the Period of the New Testament," in *Women's Bible Commentary*, eds. Carol A. Newsom and Sharon H. Ringe (Louisville, KY: Westminster John Knox, 1992), 482–488, provides a summary of the social and economic position of women in the Roman Empire during the New Testament period.

22. Karen Jo Torjesen, "Reconstruction of Women's Early Christian History," in *Searching the Scriptures: A Feminist Introduction*, Vol. 1, ed. Elizabeth Schuessler Fiorenza (New York: Crossroad, 1993), 291.

23. Thanks to Gonzaga University student Isabel Beaulieu for this observation.

24. Bruce Chilton, *Rabbi Paul: An Intellectual Biography* (New York: Doubleday, 2004), gives a fascinating description of the culture of Tarsus.

25. The NABRE offers a helpful note at Acts 13:9: "Saul, also known as Paul: there is no reason to believe that his name was changed from Saul to Paul upon his conversion. The use of a double name, one Semitic (Saul), the other Greco-Roman (Paul), is well attested (cf. Acts 1:23, Joseph Justus; Acts 12:12, 25, John Mark)."

26. David Trobisch, *Paul's Letter Collection: Tracing the Origins* (Minneapolis: Fortress, 1994), 55.

27. Dennis Ronald MacDonald, *The Legend and the Apostles: The Battle for Paul in Story and Canon* (Philadelphia: Westminster, 1983), 105–106, note 1.

28. Robert Jewett, *Paul the Apostle to America: Cultural Trends & Pauline Scholarship* (Louisville, KY: Westminster, 1994), 45–58.

29. See Susan A. Calef, "Thecla, Acts of Paul and," in *The New Interpreter's Dictionary of the Bible S–Z*, Vol. 5, ed. Katharine Doob Sakenfeld et al. (Nashville: Abingdon, 2009), 550: "That the story of Paul's virgin convert circulated independently of the larger work and is extant in multiple ancient languages attests to its popularity. . . . The Thecla story inspired a vibrant and widespread cult that lasted well into the 5th cent. and apparently also inspired some women's aspirations to ministry."

30. See Jouette M. Bassler, *1 Timothy, 2 Timothy, Titus* (Nashville: Abingdon, 1996), 17–21, for a clear presentation of the evidence against Pauline authorship of these letters. For an opposing view, see Luke Timothy Johnson, *Letters to Paul's Delegates: 1 Timothy, 2 Timothy, Titus* (Valley Forge, PA: Trinity Press International, 1996), 1–33.

31. Charles H. Talbert, *Reading Corinthians: A Literary and Theological Commentary on 1 and 2 Corinthians* (New York: Crossroad, 1992), 92–93. In 1 Corinthians 6:13a, Paul quotes his opponents; in v 13b he gives his own view. The NABRE makes this clear with quotation marks: "'Food for the stomach and the stomach for food,' but God will do away with both the one and the other." Again in 1 Corinthians 14:21–22 he quotes those with whom he disagrees; he gives his own opposing view in verses 23–25. Translations do not help the reader to make this distinction in this case. But Paul is clearly saying that speaking in tongues cannot be for unbelievers (as others in the Corinthian community have been saying, according to verse 22), because they will just think the speakers are crazy (Paul's view stated in verse 23). Prophecy, on the other hand, is not only for believers (verse 22) but for unbelievers, too (verses 24–25). See also Stark, *The Rise of Christianity*, 108. Stark reports that Laurence Iannaccone made the same argument about 1 Corinthians 14:33–36 in 1982.

32. Torjesen, "Reconstruction of Women's Early Christian History," in *Searching the Scriptures*, Vol. 1, ed. Schuessler Fiorenza, 291, indicates "that the theological

foundations for women's leadership were laid in the Christian vision of the reign of God (*basileia*) in which God's wholeness was extended to all, making women equal to men." See 293 regarding women in position of leadership.

33. In 1981 the theologian Karl Rahner asserted the appropriateness and duty of Roman Catholic theologians to examine respectfully but critically such official declarations. In his essay "Women and the Priesthood," in *The Content of Faith: The Best of Karl Rahner's Theological Writings* (New York: Crossroad, 1992), 424–433, Rahner points out that it takes time for the Church to fully register the social significance of an abstract moral principle like, in Christ "[t]here is neither Jew nor Greek, there is neither slave nor free person, there is not male and female" (Galatians 3:28). The tide of social custom was strongly against the abolition of slavery; but the time eventually came for the Church to see that slavery was not tolerable, despite its ubiquitous and unquestioned practice in the past. In the same way the time has been emerging to eliminate the subordination of women to men in Church and society, whether that means ordaining women as deacons or priests or not.

34. Phyllis Zagano, *Holy Saturday: An Argument for the Restoration of the Female Diaconate in the Catholic Church* (New York: Crossroad, 2000).

35. Phyllis Zagano, "A Woman on the Altar: Can the Church Ordain Women Deacons?" in *U.S. Catholic*, November 2011.

36. See also Stark, *The Rise of Christianity*, 107–110. Stark has investigated gender and religious roles in the first five centuries beginning before Christianity became the dominant religion of the empire. During those five centuries women made up a higher proportion of Christians than men. Under such circumstances, women typically enjoy greater freedom and power.

Stark calls attention to 1 Timothy 3:11, which the King James Version translated, "Even so must their wives (*gynaikas*) be grave," in contrast to modern translations such as the NABRE: "Women (*gynaikas*), similarly, should be dignified." The NABRE note says: "[T]his seems to refer to women deacons but may possibly mean wives of deacons. The former is preferred because the word is used absolutely; if deacons' wives were meant, a possessive 'their' would be expected. Moreover, they are also introduced by the word 'similarly,' as in 1 Timothy 3:8; this parallel suggests that they too exercised ecclesiastical functions."

In 1 Timothy the term "deacon" designates an office, not simply a function. James D. G. Dunn, *The First and Second Letters to Timothy and the Letter to Titus: Introduction, Commentary, and Reflections* (Nashville: Abingdon, 2000), 807, reports that the Roman governor Pliny, writing in 112 CE, two or three decades after 1 Timothy, speaks of two Christian women as "maids who were called ministers/deacons" (*Letters* 10.96). That also sounds like an office, not a function. While 1 Timothy 3 enumerates character traits for deacons as well as bishops, nothing is said about responsibilities of deacons in the Church. The passage 1 Timothy 2:11–12 possibly suggests that a female deacon was not to teach or have authority over a man. If so this may be the view of the author of the letter writing under the authority of Paul's name; it may be that this letter writer is asserting his opposition to current practice that non-Christian neighbors found distressing because it violated the social norms.

## 5. VIOLENCE AND DEATH

1. Mark 13:26 speaks of the Son of Man coming "in" (*en*) the clouds; Mark 14:62 says "with" (*meta*) the clouds. For a discussion of the various texts and translations of Daniel 7:13, see Marcus, *Mark 8–16*, 905.

2. Herman Waetjen, *The Origin and Destiny of Humanness: An Interpretation of the Gospel According to Matthew* (Corte Madera, CA: Omega, 1976), states in the foreword: "An effort has been made to avoid sexist language and to speak as inclusively as possible. . . . The phrases 'the Son of the Human Being' and 'the Human Being' are substituted for 'the Son of Man.'"

3. Lohfink, *Jesus of Nazareth*, 67.

4. Frederick Danker, ed., *A Greek-English Lexicon of the New Testament and Other Early Christian Literature* (Chicago: University of Chicago Press, 2000), 507: "Not a toponym *from Cana* (Jerome) nor *Canaanite,* but fr. Aram. קַנְאָן [*qanan*] 'enthusiast, zealot' (cp. Lk 6:15; Ac 1:13, where he is called ζηλωτής [*zelotes*]), prob. because he had formerly belonged to the party of the 'Zealots' or 'Freedom Fighters.'"

5. There is much confusion among the evangelists about the actual names of the Twelve. Levi the tax collector is not included among the Twelve in Mark 4:16–19. Instead of Levi the son of Alphaeus, Mark includes James the son of Alphaeus. Mark includes someone he calls Matthew in his list of the Twelve. The Evangelist Matthew adds "the tax collector" after this name in Matthew 10:3. Matthew also substitutes the name Matthew for Levi in his retelling of Mark's story of the call of Levi (Matthew 9:9; Mark 2:14). The Evangelist Matthew was clearly puzzled by what he heard in Mark's story and solved the puzzle by changing Levi's name to Matthew, clearly making Mark's Levi one of the Twelve.

6. Marcus, *Mark 1–8*, 465.

7. Mark's description of the woman first as a "Greek" and then as a "Syrophoenician" would have been an emotional bombshell for Mark's listeners. Both 1 and 2 Maccabees perpetuate horror stories of what the Greek ruler in Syria, Antiochus IV, did to Israelites

just two centuries earlier. He profaned the Temple; outlawed the Sabbath, circumcision, and the traditional feasts; and forced Israelites to abandon the law and participate in Greek sacrifices. Those who did not comply were tortured to death; 2 Maccabees 6–7 provides a vivid account of this torture. This was the cause of the Maccabean Revolt, which was successful and had served as the inspiration for the violent revolt against Rome around the time Mark was telling his story of Jesus. Syria had continued to be a source of hostile provocation in the Roman era. The census-taking by the Syrian governor Quirinius in 6 BCE and reported in Luke 2:1–2 and Acts 5:37 had provoked a widespread revolt in Galilee. David Rhoads reports that to make matters worse, in Mark's time, near the time of the Great Revolt, the people of Tyre in fact assaulted Israelites in their city. The people of Tyre got their food from Galilee, their breadbasket, and ate well while the Galileans had barely enough for themselves. The woman is depicted as one of those wealthy Tyrians. Her daughter is lying on a bed (*klinen*), not a peasant's mat (*krabbatos*, Mark 2:4, 9, 12). David Rhoads, *Reading Mark: Engaging the Gospel* (Minneapolis: Fortress, 2004) 93, 232, n. 41.

8. The identity of the four thousand is disputed. Marcus, *Mark 1–8*, 490, reviews the way some have identified the four thousand as Gentiles on the basis of the number. They reduce the four thousand to four, which Marcus says is not implausible, and then propose that the number four refers to the four corners of the earth or the four winds. Marcus does not see this as convincing. The saying in Q concerning people coming from the four corners of the earth (Matthew 8:11–12; Luke 13:28–29) seems more to be about the regathering of scattered Israel. He concludes therefore that the four thousand are more likely Israelites. David Rhoads, however, clearly identifies them as Gentiles in *Reading Mark*, 85, 87. This conclusion is supported not by the number four thousand but by the narrative sequence in Mark's story. Gerhard Lohfink concurs that Jesus goes into Gentile territory in Mark. He thinks that the historical Jesus never left Israelite territory except possibly to seek out Israelites living in marginal areas, but this is in contrast to the picture Mark presents. See Lohfink, *Jesus of Nazareth*, 68.

9. Like so much in Mark's gospel, it is easy to miss the pattern of inclusion, first of alienated Israelites, then of Gentiles, as a deliberate movement in fulfillment of Israel's world-transforming vocation. In conversation, Tom Boomershine called attention to this pattern and its culmination in Jesus's cleansing of the Temple as a "house of prayer *for all peoples*." The last three words are peculiar to Mark, omitted in Matthew's and Luke's retelling of Mark's story.

10. In his mission to the Gentiles, Jesus once again is not doing something innovative. His vision is rooted in his people's own traditions.

11. Josephus, *Jewish Antiquities* XI, 325–339.

12. Adela Yarbro Collins, *Mark* (Minneapolis: Augsburg Fortress, 2007), 515.

13. When Matthew repeats Mark's story, he augments it by actually quoting Zechariah rather than alluding to it.

14. Ben C. Ollenburger, "The Book of Zechariah: Introduction, Commentary, and Reflections," in *The New Interpreter's Dictionary of the Bible S–Z*, 807.

15. Collins, *Mark*, 515.

16. When Jesus enters the city in Mark's story he simply looks around; then, because it is late, he leaves. In Matthew and Luke, Jesus proceeds immediately to the Temple to cleanse it of what has defiled it, as Judas Maccabeus does in 1 Maccabees 4:36–37. Marcus, *Mark 8–16*, 780, remarks that this is how a Messiah is supposed to act. Collins, *Mark*, 521, suggests that Jesus's odd delay of taking action in Mark could well have frustrated the listeners who expected some kind of great thing to happen, either further celebration, or some act of cleansing as in the stories in 1 Maccabees.

17. Both Rudolf Schnackenburg, *Das Evangelium Nach Markus* (Dusseldorf: Patmos, 1971), 134, and Marcus, *Mark 8–16*, 793, view Jesus's action as taking place in the Court of the Gentiles. Rudolf Pesch, *Das Markusevangelium II Teil: Kommentar Zu Kap. 8, 27–16, 20* (Freiburg: Herder, 1977), 197, concurs, observing, "From the Mount of Olives a person would come into the Temple from the East Gate into the Court of the Gentiles" (Adam Bartholomew translation).

Jesus's actions of driving out those selling and buying and overturning the tables of the money changers and the seats of those selling pigeons raise the question of Jesus's own propensity to violence. In John's version of the story, placed at the beginning of his gospel instead of toward the end, Jesus's violent actions are escalated to include making a whip of cords and driving out animals as well as humans. Luke, on the other hand, reduces the description to Jesus's driving out those who engaged in commerce. We should at least note that in no version of the story does Jesus physically harm either animal or human. This is the only action that could qualify as violent, and the violence is very limited. If here we see Jesus as acceding to the use of violence, it is a singular instance executed against a background of consistent rejection of it. If anyone is inclined to take it as authorizing violence on the part of his followers, one could only hope that those followers would keep the damage they do within the limits Jesus modeled.

18. Walter Wink, *Engaging the Powers: Discernment and Resistance in a World of Domination* (Minneapolis: Fortress, 1992), 244.

19. David Rhoads, *Israel in Revolution: 6–74 C.E.* (Philadelphia: Fortress, 1976), 91. See Josephus, *Jewish War II*, 409, 411–412 in *Josephus in Nine Volumes*, trans. H. St. J. Thackery, Vol. II (Cambridge, MA: Harvard University Press, 1927):

> Eleazar . . . persuaded those who officiated in the Temple services to accept no gift or sacrifice from a foreigner. This action laid the foundation of the war with the Romans. . . . Thereupon

> the principal citizens assembled with the chief priests and the most notable Pharisees to deliberate on the position of affairs, now that they were faced with what seemed irreparable disaster. Deciding to try the effect of an appeal to the revolutionaries, they called the people together. . . . They began by expressing the keenest indignation at the audacity of this revolt and at their country being thus threatened by so serious a war. Then they proceeded to expose the absurdity of the alleged pretext. Their forefathers, they said, had adorned the sanctuary mainly at the expense of aliens and had always accepted the gifts of foreign nations.

20. N. T. Wright, *Paul and the Faithfulness of God* (Minneapolis: Fortress, 2013), 83.

21. Jesus's action in the Temple is bracketed by the story of Jesus's cursing of the fig tree and the discovery that it has withered (11:12–14, 19–25). This strange story is most likely prophetically symbolic of something involving the Temple and its leadership. Whatever the reason for and details involved in the action of the historical Jesus in the Temple—and the different evangelists tell the story differently and suggest different meanings—the vocabulary associated with the revolutionaries who occupied the Temple and provoked the Great Revolt that brought the Temple to ruins must have made that disaster uppermost in the minds of Mark's listeners.

22. Aquilina and Papandrea, *Seven Revolutions: How Christianity Changed the World and Can Change It Again*.

23. Philip Schaff, *History of the Christian Church, Vol. II: Ante-Nicene Christianity. A.D. 100–325* (New York: Charles Scribner's Sons, 1914), 75: "To these protracted and cruel persecutions the church opposed no revolutionary violence, nor carnal resistance, but the moral heroism of suffering and dying for the truth. . . . In this very heroism she proved worthy of her divine founder, who submitted to the death of the cross for the salvation of the world, and even prayed that his murderers might be forgiven." William T. Cavanaugh, *Field Hospital: The Church's Engagement with a Wounded World* (Grand Rapids, MI: William B. Eerdmans, 2016), 172: a theology of martyrdom was based on "the fact that the martyrs chose to receive death rather than to deal it out. They did so as an act of *imitatio Christi*, not in the hope that more blood would eventually be shed, but in the conviction that the triumph of Christ through his death and resurrection meant that the cycle of violence had been broken."

24. See David G. Hunter, "A Decade of Research on Early Christians and Military Service," in *Religious Studies Review* 18, 2 (1992): 87–94.

25. Peter J. Leithart, *Defending Constantine: The Twilight of an Empire and the Dawn of Christendom* (Downers Grove, IL: IVP Academic, 2010), 259.

26. When it is not capitalized, "orthodox" is a descriptive adjective meaning "right thinking." It does not designate the Eastern Orthodox Churches. As the Church spread during the first few centuries, varieties of beliefs developed that led to rich debate. The

term "orthodox" is used to identify those Christians who were prepared to formulate the doctrinal creeds of the Ecumenical Councils in which bishops throughout the Roman Empire participated. The first of these was called at Nicea by the Emperor Constantine in the fourth century. He hoped to promote uniformity in the Church that would in turn aid in unifying the Empire. The term "orthodox" distinguished these Christians from Christians who did not agree with those creeds. The ongoing debates of the previous centuries laid the groundwork for the agreements hammered out in the councils.

The result of these early debates was that Christians began to group together around a variety of shared beliefs, forming a variety of groups. Scholars label as "proto-orthodox" the group whose views won out in the great councils. Hurtado, *Lord Jesus Christ: Devotion to Jesus in Earliest Christianity*, 10–12, lists four characteristics that connected the Christians whom scholars label "proto-orthodox": "a readiness to recognize one another (despite differences), . . . a high regard for traditions and suspicion of radical innovations, . . . a commitment to the 'Old Testament' writings as scripture, and . . . an exclusivist 'monotheistic' stance in which the deity of the Old Testament is the only valid deity worthy of worship."

27. The Q tradition incorporated by Matthew and Luke includes the instruction to turn the other cheek when struck and give the second garment when sued or robbed (Matthew 5:38–41; Luke 6:29). Wink, *Engaging the Powers*, 175–193, shows how this is not counsel to be passive but to resist without becoming violent. Jesus has in mind someone in a position of superiority backhanding a person beneath that person. To turn the other cheek is to challenge that person to hit again, but with the fist, which is the way one hits an equal. The person is without violence asserting her or his dignity. Stripping off the second garment and giving it to the one who sues or robs you would, in Jesus's culture, bring shame not on the one who is naked but on the one looking on another's nakedness (Genesis 9:20–25). It is clear that the early Christians were not passive. They refused to sacrifice to the emperor, for example. That is a form of active resistance.

28. Cavanaugh, *Field Hospital*, 160.

29. Walter Kasper, *Mercy: The Essence of the Gospel and the Key to Christian Life* (New York: Paulist, 2014), 192.

30. J. Philip Wogaman notes in *Christian Ethics: A Historical Introduction* (Louisville, KY: Westminster John Knox, 1993), 114, that Martin Luther "distinguished between personal self-defense, which he rejected as contrary to Christian love, and the defense of others, which he considered to be an expression of that same love."

31. Stephen Jay Gould, "A Time of Gifts," *New York Times*, September 26, 2001.

32. The rejection of a belief in a literal Paradise at the beginning of human history has been expressed in recent times also by Teilhard de Chardin and Karl Rahner; it can be found already in the seventh century in the writings of Maximus the Confessor.

See Elphinstone, *Freedom, Suffering and Love*, 6. For Teilhard de Chardin, see Delio, *The Unbearable Wholeness of Being*, 116–117; for Rahner's rejection of Paradise as a past reality, see Cannato, *Field of Compassion*, 58, 71; for Maximus's view, see Hans Urs von Balthasar, *Cosmic Liturgy* (San Francisco: Ignatius Press, 2003), 133. See also Ivone Gebara, *Longing for Running Water: Ecofeminism and Liberation* (Minneapolis: Augsburg, 1999), 96–99.

This view that humans were originally a violent species that must evolve toward love is not accepted by all. Marc Ian Barasch, *Field Notes on the Compassionate Life: A Search for the Soul of Kindness* (Rodale, 2005), 27–28, does not agree that from the beginning, we were violent and lacked the capacity for love. See also Erich Fromm, *The Anatomy of Human Destructiveness* (New York: Holt, Rinehart and Winston, 1973), 124–129, for paleontological evidence that our hominid ancestors were omnivorous and lacked the instinctual and morphological equipment found in carnivorous predators like wolves. Consequently they cannot be viewed as a paleontological "Adam" responsible for human destructiveness.

Though the jury may still be out on this question of when the capacity for love enters the history of our species, advocates of both views agree that as a species we have used our intelligence in the service of destruction and that the survival of life on our planet requires that we learn to love. It is also important to affirm the possibility of adapting our theological story to the new story being told by science when necessary.

## 6. VISIONARIES, PROPHETS, AND SAINTS

1. Antonio Spadaro, S.J., "A Big Heart Open to God: An Interview with Pope Francis," *America Magazine*, September 30, 2013.

2. Pierre Teilhard de Chardin, trans. Rene Hague, *The Making of a Mind: Letters from a Soldier-Priest 1914–1918* (New York: Harper & Row, 1965), 119–120.

3. John Grim and Mary Evelyn Tucker, *Teilhard de Chardin: Biography*, American Teilhard Association website, http://teilharddechardin.org/index.php.

4. *Letter to Christophe de Gaudefroy, 7 October 1929, Lettres inédites,* 80, American Teilhard Association website, http://teilharddechardin.org/index.php/teilhards-quotes.

5. Mary Gilbert and Thomas King, eds., *Letters of Teilhard de Chardin and Lucile Swan* (University of Scranton Press, 2005).

6. Stephen Jay Gould, *Hen's Teeth and Horse's Toes: Further Reflections in Natural History* (New York: W. W. Norton, 2010), 226.

7. Beatrice Bruteau was one of Teilhard's most brilliant students and the cofounder of the American Teilhard Association. She was nevertheless critical of his "xenopho-

bic Catholicism." In light of the scientific developments in astrophysics and cellular biology since Teilhard's death, she confirmed and updated his basic view. In response to Teilhard's view of the Omega point as a final union of all things in Christ, she advocated for the Trinitarian God as both the source and the future of a universe that evolves in the direction of open-ended expansion. See Cynthia Bourgeault, "Teilhard, the Trinity, and Evolution." In *Personal Transformation and a New Creation: The Spiritual Revolution of Beatrice Bruteau*, ed. Ilia Delio (Maryknoll, NY: Orbis, 2016), 75–76, 79–80.

8. Juergen Moltmann takes Teilhard de Chardin to task for articulating a vision of a future in which the universe incarnates Christ without redeeming all the victims of this violent process. Commenting on his thoughts about Christ and the universe in *The Way of Jesus Christ* (New York: HarperCollins, 1990), 296, Moltmann says, "A *Christus evolutor* without *Christus redemptor* is nothing other than a cruel *Christus selector*, a historical world-judge without compassion for the weak, and a breeder of life uninterested in the victims." Moltmann continues in terms specific to the way science describes the world's evolution: "There is certainly a history of ongoing creation; there are evolutions to richer and more complex forms of life. . . . But in this history of creation there is also dying, violent death, mass extermination and the extinction of whole species through natural catastrophes and epidemics."

Science cannot picture a future for what evolution has left behind. Only theology can do that. Employing the difficult theological term "eschatology," which derives from the Greek word meaning what is "farthest" or "last," Moltmann says,

> What is eschatological is the new creation of all things which were and are and will be. What is eschatological is the bringing back of all things out of their past, and the gathering of them into the kingdom of glory. What is eschatological is the raising of the body and the whole of nature. What is eschatological is that eternity of the new creation which all things in time will experience simultaneously when time ends. To put it simply: God forgets nothing that he has created. Nothing is lost to him. He will restore it all. (303)

See also Tollefsen's "Christocentric Cosmology," in *The Oxford Handbook of Maximus the Confessor*, 319; Paul M. Blowers, "Unfinished Creative Business: Maximus the Confessor, Evolutionary Theodicy, and Human Stewardship in Creation," in *On Earth as It Is in Heaven: Cultivating a Contemporary Theology of Creation* (Grand Rapids, MI: William B. Eerdmans, 2016), 181, 186.

9. Thomas Berry, *Teilhard de Chardin and the Age of Ecology*, interview with Jane Blewett, Thomas Berry Foundation, from the Library of Lou Niznik, published October 5, 2013. http://thomasberry.org/publications-and-media/teilhard-de-chardin-in-the-age-of-ecology-by-thomas-berry.

10. Susannah Heschel, *The Aryan Jesus: Christian Theologians and the Bible in Nazi Germany* (Princeton, NJ: Princeton University Press, 2008), 16.

11. Rabbi Marc H. Tanenbaum, "Heschel and Vatican II: A Memorial Symposium in Honor of Rabbi Abraham Joshua Heschel," *Archives of the American Jewish Committee* (New York, February 21, 1983).

12. Abraham Joshua Heschel, *The Prophets* (New York: Harper & Row, 1962), 223–224.

13. Susannah Heschel, "What Does It Mean to Be Religious in an Age of Neoliberalism?" *Faith and Leadership*, November 3, 2014.

14. Nadine Epstein, "Susannah Heschel on the Legacy of Her Father, Rabbi Abraham Joshua Heschel, and the Civil Rights Movement," *Moment* Magazine, April 30, 2015.

15. Susannah Heschel, "Following in My Father's Footsteps," *Vox of Dartmouth*, April 4, 2005.

16. Robert M. Dowling, *Critical Companion to Eugene O'Neill: A Literary Reference to His Life and Work* (New York: Facts on File, 2009), 563.

17. *Catholic Worker*, January 1942, Volume 1, Issue 4.

18. Kate Hennessy, *Dorothy Day: The World Will Be Saved by Beauty* (New York: Scribner, 2017), 168–205.

19. Shannon Hill, "Dorothy Day: A Sinner and Nearly a Saint," *Publishers Weekly*, January 3, 2017.

20. Bob Fitch Photo Archive, Stanford University Libraries.

## 7. THE SEED AND THE SOIL

1. Virgil F. Thompson, *The New Testament in Contemporary Life* (Dubuque, IA: Kendall Hunt, 2011), 65.

2. Mary Ann Tolbert, *Sowing the Gospel: Mark's World in Literary and Historical Perspective* (Minneapolis: Fortress, 1989), 153: "[E]ach type of response, each type of ground, is from the outset clearly identified as a *group of people*. . . . The contrast should suggest to the audience that 'these' are groups nearby, recently discussed; 'others' and 'those' are not yet fully explicated."

3. Daniel is distressed and terrified by the vision and asks someone present for an explanation. Among other things, the interpreter identifies the "one like a son of man" with "the holy ones of the Most High" (Daniel 7:18, 21–22, 25, 27). Although these are often members of the heavenly court or angels, the fact that in 7:21 they are attacked and

conquered by a horn of the fourth beast makes it more likely that these are the people of Israel. Two consequences follow from this. First, the term is not masculine but generic, since it designates all Israelites. So it is now more often translated "Child of Humanity" or "the Human One." We retain "Son of Man" as our translation of Mark's term in order to preserve the allusion to the vision in Daniel 7, where most standard translations continue to translate it that way. The second consequence is that in the gospels Jesus sees himself as bringing to fulfillment the role envisioned for his people. This does not mean he appropriates it exclusively for himself. His mission is to regroup Israel, represented by his circle of twelve, who in turn represent the twelve tribes regathered, so they will fulfill their vocation of testifying to their God before the nations. (See Ezekiel 36:16–36 for the connection between the regathering of the scattered tribes of Israel, a new heart of obedience to the Torah, the restoration of the land, and the testimony to God's holiness among the nations.)

4. The identification of Jesus as "Son of Man" in the New Testament is intriguing. With the exception of Stephen's words in Acts 7:56 just before he is stoned, this phrase is found only on the lips of Jesus. While scholars debate whether this identification goes back to Jesus himself, Mark develops his story in a way that both makes the identification unmistakable and yields a narrative about the Son of Man for which the vision in Daniel 7 provides the climax, but is played out in the life of Jesus. This results in Mark in a narrative in three stages: earthly ministry, suffering, and returning in triumph.

5. The Greek word is *skandalisthesesthe*, which connotes an internal feeling of offense. See Thomas E. Boomershine, *The Messiah of Peace: A Performance-Criticism Commentary on Mark's Passion-Resurrection Narrative* (Eugene, OR: Cascade Books, 2015), 103, for an analysis of Mark's use of this word.

6. Cannato, *Field of Compassion*, 81–95. Jesus's passion is for justice and compassion, not personal holiness and purity.

7. Thomas Berry, *Selected Writings on the Earth Community* (Maryknoll, NY: Orbis, 2014), 1, 4, 17.

8. Thomas Berry, *The Christian Future and the Fate of the Earth* (Maryknoll, NY: Orbis, 2009), 28–29, 31, 35–40.

9. Berry, *Selected Writings on the Earth Community*, 6–7.

10. Ibrahim Ozdemir, "Toward an Understanding of Environmental Ethics from a Qur'anic Perspective," in *Islam and Ecology: A Bestowed Trust*, eds. Frederick M. Denny, Richard C. Foltz, and Azizan Baharuddin (Cambridge, MA: Harvard University Press, 2003), 5.

11. Berry, *The Christian Future and the Fate of the Earth*, xiii, 30.

12. Berry, *Selected Writings on the Earth Community*, 7–8, 11, 13.

13. Berry, *The Christian Future and the Fate of the Earth*, xxiii.

14. Berry, *Selected Writings on the Earth Community*, 18.

15. See also Isaiah 11:1–9; 40:3–4; Hosea 2:21–25.

16. Ivone Gebara, *Longing for Running Water: Ecofeminism and Liberation* (Minneapolis: Augsburg, 1999), 27–28.

17. Elizabeth A. Johnson, *Women, Earth, and Creator Spirit* (Mahwah, NJ: Paulist, 1993), 10.

18. Johnson, *Women, Earth, and Creator Spirit*, 13.

19. N. T. Wright, *How God Became King* (New York: HarperOne, 2012), 192.

20. Berry, *Selected Writings on the Earth Community*, 16.

21. Lohfink, *Jesus of Nazareth*, 44–46.

22. Bauckham, *The Bible and Ecology*, 167: "The activities of Jesus were small-scale anticipations of the Kingdom that heralded its universal coming in the future. What is notable about them . . . is the way that their holistic character points to the coming of the Kingdom in all creation."

23. C. H. Dodd, *The Parables of the Kingdom* (New York: Charles Scribner's Sons, 1961), 5.

## 8. HERE AND NOW

1. A common understanding of salvation is that the eternal life of the individual soul achieves unity with the Creator in a spiritual realm after death. And yet Christianity had a revolutionary impact on how ordinary people would live together and flourish in this world. As early Christians looked to the Bible to guide their lives, they could not miss the fact that it was filled with instructions about the right way to live in the present. But the significance of earthly living was eventually reinterpreted by Platonic philosophy's influence on Christianity. No longer were spirit-endowed creatures to find fulfillment by participating in the evolution of the cosmos that would culminate in the restoration of Paradise. Instead, earthly life would be a proving ground for a totally spiritual life in a non-material realm. At the beginning of the Italian Renaissance, when interest in and a new capacity for improving the quality of life in this world for its own sake resurfaced, Christians had a tradition of earthly living just waiting to be employed to that end.

2. Peter J. Leithart, *The Four: A Survey of the Gospels* (Moscow, ID: Canonpress, 2010), 13.

3. The question of the role of Gentiles and Israelites in the spread of the Christian movement, and of Gentile philosophy and its story of the world in the interpretation of

Israel's traditions, is complex and becoming more fascinating as scholars become more consciously aware of its complexity. According to Rodney Stark, *The Rise of Christianity*, chapter 3, "The Mission to the Jews: Why It Probably Succeeded," 49–71, Hellenistic Israelites continued to be a significant source of converts to the Christian sect at least as late as the fourth century. Like Philo of Alexandria, these converts surely had already made a significant accommodation to Hellenistic culture.

4. N. T. Wright, *Paul and the Faithfulness of God: Christian Origins and the Question of God*, Vol. 4 (Minneapolis: Fortress, 2013), 209:

> [Plato] exercised a massive influence over the next thousand years of western thought, not least some key elements in early Christianity. . . . In particular . . . Plato taught that the world of space, time and matter was essentially a secondary thing, a world of illusion, by comparison with the ultimate reality, the world of the "Forms" of "Ideas", the invisible realities of which this-worldly things (whether trees and chairs, or instance of good behavior) were mere space-time copies. True "knowledge" was therefore knowledge of the Forms. . . . This "knowledge" was to be the main goal and occupation, not of the outward bodily senses, but of the soul, which, Plato believed, was immortal, coming into a human body and passing from it either to a state of disembodied bliss or into a sequence of other bodies, through reincarnation.

Diogenes Allen and Eric O. Springsted, *Philosophy for Understanding Theology* (Louisville, KY: Westminster John Knox, 2007): "Plato's various views on the nature of the soul caused Augustine and other Christian Platonists difficulties. They were actually more immediately in contact with the Platonic revival and especially Plotinus than with Plato himself. . . . But the Platonic tradition that deeply influenced many theologians led them to think of the human soul as a spiritual substance that could exist disembodied" (35). "We come now to Plotinus himself (AD 205–270). . . . Plotinus draws mostly on those dialogues of Plato which stress that our proper life is to be found by a knowledge of another realm (the *Phaedo, Phaedrus,* and the *Symposium*) or on those parts of the *Timaeus* and the *Republic* which do the same. As for other Platonists, so for Plotinus the soul is divine and the object of life is to understand how we may restore the soul to its proper place" (49).

N. T. Wright, *Surprised by Hope: Rethinking Heaven, the Resurrection, and the Mission of the Church* (New York: HarperOne, 2008), 158, identifies leading theologians in the post–New Testament patristic and medieval periods who clearly believed in a resurrection of the body following a period of disembodiment between the death and resurrection. This belief was rooted in a strong view that God's material creation was good and must not be abandoned. But much medieval piety was based on a belief that the dead were destined for a non-earthly heaven or hell. If people continued to speak of resurrection at all, it "seemed simply to be a rather special way of talking about heaven."

5. Philo, *On Dreams* I, 256, in *Philo in Ten Volumes (and Two Supplementary Volumes)*, trans. F. H. Colson and G. H. Whitaker, Vol. V (Cambridge: Harvard University Press, 1968): "For so shalt thou be able also to return to thy father's house, and be quit of that long endless distress which besets thee in a foreign land." Philo employs the phrase "thy father's house," which is a reference to heaven, the destiny of the soul. In contrast, "a foreign land" signifies the earthly realm. See N. T. Wright, *The Resurrection of the Son of God* (Minneapolis: Fortress, 2003), 446, n. 134. See also Philo, *On the Confusion of Tongues*, 77–78, in *Philo in Ten Volumes*, trans. Colson and Whitaker, Vol. IV:

> This is why all whom Moses calls wise are represented as sojourners. Their souls are never colonists leaving heaven for a new home. Their way is to visit earthly nature as men who travel abroad to see and learn. So when they have stayed awhile in their bodies, and beheld through them all that sense and mortality has to shew, they make their way back to the place from which they set out at the first. To them the heavenly region, where their citizenship lies, is their native land; the earthly region in which they became sojourners is a foreign country.

6. Moltmann, *The Coming of God: Christian Eschatology*, 268. Moltmann passes on the conclusion of Paul Althaus that "Transformation, not annihilation—that is the unanimously held doctrine from Irenaeus onward, by way of Augustine and Gregory the Great, Aquinas and the whole of medieval theology, down to present-day Catholic dogmatics." It was Lutheran orthodoxy that for one hundred years following the second dispute about the Eucharist universally espoused the view that the ultimate fate of the earth was not transformation but annihilation. It appears, however, that the transformation of the earth had little impact on the popular view that eternal life would be the life of the disembodied soul in a realm other than this earth.

7. Collins, *The Scepter and the Star*, 1. He quotes the Jewish scholar Gershom Scholem:

> Judaism "has always maintained a concept of redemption as an event which takes place publicly, on the stage of history, and within the community." Christianity, in contrast, locates redemption in "the spiritual and unseen realm, . . . in the private world of each individual."

This accurately describes the popular perception. Our study has been aimed at demonstrating that Mark shares what Scholem describes as the Jewish view.

8. Stephen Prothero, *American Jesus: How the Son of God Became a National Icon* (New York: Farrar, Straus and Giroux, 2003), 19–42.

9. N. T. Wright, *Paul and the Faithfulness of God*, 163–164.

10. Heidi Ann Russell, *Quantum Shift: Theological and Pastoral Implications of Contemporary Developments in Science* (Collegeville, MN: Liturgical Press, 2015), 77:

> Rarely in our congregations do we see the same emphasis on care for creation that we see on doing Bible studies, joining prayer groups, or taking our kids to Sunday School. Congregations do not overwhelmingly seem to associate a concern for the entirety of creation with the heart of Gospel values and what it means to be a Christian. Particularly in a consumer culture, there is a sense that the earth exists to serve us rather than we exist to serve creation. Environmental concerns are secondary to economic concerns. The well-being of humanity, not only in the sense of survival but also in the sense of property and wealth, is put before concern for the system as a whole.

11. Brian D. McLaren, *We Make the Road by Walking: A Year-Long Quest for Spiritual Formation, Reorientation, and Activation* (New York: Jericho Books, 2014), 261.

# Bibliography

## PRIMARY SOURCES

*The Book of Common Prayer*. New York: The Church Pension Fund, 1945.

Dupont-Sommer, Andrei. *The Essene Writings from Qumran*. Cleveland: Meridian, 1961.

*Early Christian Fathers*. Edited by Cyril C. Richardson with John T. McNeill, John Baillie, and Henry P. Van Dusen. Vol. 1, *The Library of Christian Classics*. New York: Macmillan, 1970.

Eusebius. "Church History." In *A Select Library of Nicene and Post-Nicene Fathers of the Christian Church, Translated into English with Prolegomena and Explanatory Notes*. Second Series. Vol. I. Edited by Philip Schaff and Henry Wace. New York: The Christian Literature Company, 1890.

*A Greek-English Lexicon of the New Testament and Other Early Christian Literature*. [*Griechisch-Englisches Woerterbuch zu den Schriften des Neuen Testaments und der fruehchristlichen Literatur*, 6th ed. Edited by Kurt and Barbara Aland with Viktor Reichmann.] Translated by William F. Arndt and F. Wilbur Gingrich. 3rd ed. Revised and edited by Frederick Danker. Chicago: University of Chicago, 2000.

*The HarperCollins Study Bible: New Revised Standard Version*. Edited by Wayne A. Meeks et al. 1993.

Josephus. "Jewish Antiquities, Books IX–XI." Translated by Ralph Marcus. In *Josephus in Nine Volumes*. Vol. VI. The Loeb Classical Library. Cambridge: Harvard University Press, 1937.

Josephus. "Jewish Antiquities, Books XV–XVII." Translated by Ralph Marcus and Allen Wikgren. In *Josephus in Nine Volumes*. Vol. VIII. The Loeb Classical Library. Cambridge: Harvard University Press, 1963.

Josephus. "Jewish Antiquities, Books XVIII–XX. General Index to Volumes I–IX." Translated by Louis H. Feldman. In *Josephus in Nine Volumes*. Vol. IX. The Loeb Classical Library. Cambridge: Harvard University Press, 1965.

Josephus. "The Jewish War, Books I–III." Translated by H. St. J. Thackery. In *Josephus in Nine Volumes*. Vol. II. Loeb Classical Library. Cambridge: Harvard University Press, 1927.

Josephus. "The Jewish War, Books IV–VII." Translated by H. St. J. Thackery. In *Josephus in Nine Volumes*. Vol. III. Loeb Classical Library. Cambridge: Harvard University Press, 1928.

"Life of Adam and Eve." Translated by M. D. Johnson. In *The Old Testament Pseudepigrapha*. Vol. 2. Edited by James H. Charlesworth. Garden City, NY: Doubleday, 1985.

Luther, Martin. *Commentary on Romans*. Translated by J. Theodore Mueller. Grand Rapids, MI: Zondervan, 1954.

Maximus the Confessor. *On Difficulties in the Church Fathers. The Ambigua*. Edited by Nicholas Constas. Cambridge: Harvard University Press, 2014.

*New American Bible, Revised Edition*. Washington, DC: Confraternity of Christian Doctrine, 2010.

*New English Translation*. Biblical Studies Press, L.L.C., 1996–2005.

*A New English Translation of the Septuagint*. Edited by Albert Pietersma and Benjamin G. Wright. International Organization for Septuagint and Cognate Studies, Inc., 2014.

*The New Oxford Annotated Bible with the Apocryphal/Deuterocanonical Books*. Edited by Michael D. Coogan et al. New York: Oxford University Press, 2001.

*The Oxford Classical Dictionary*, edited by H. H. Scullard and N. G. L. Hammond. Oxford: Clarendon, 1970.

Philo. "On the Confusion of Tongues (*De Confusione Linguarum*)." Translated by F. H. Colson and G. H. Whitaker. In *Philo in Ten Volumes (and Two Supplementary Volumes)*. Vol. IV. Loeb Classical Library. Cambridge: Harvard University Press, 1932.

Philo. "On Dreams (*De Somniis*)." Translated by F. H. Colson and G. H. Whitaker. In *Philo in Ten Volumes (and Two Supplementary Volumes)*. Vol. V. Loeb Classical Library. Cambridge: Harvard University Press, 1934.

Philo. "On the Special Laws (*De Specialibus Legibus*)." Translated by F. H. Colson. In *Philo in Ten Volumes (and Two Supplementary Volumes)*. Vol. VII. Loeb Classical Library. Cambridge: Harvard University Press, 1937.

*Songs and Carols from a Manuscript in the British Museum of the Fifteenth Century.* Edited by Thomas Wright. London: T. Richards, 1856.

*The Study Quran: A New Translation and Commentary.* Edited by Seyyed Hossein Nasr. New York: HarperOne, 2015.

## SECONDARY SOURCES

Allen, Diogenes, and Eric O. Springsted. *Philosophy for Understanding Theology*, 2nd ed. Louisville, KY: Westminster John Knox, 2007.

Aquilina, Michael, and James L. Papandrea. *Seven Revolutions: How Christianity Changed the World and Can Change It Again.* New York: Image, 2015.

Auerbach, Erich. *Mimesis: The Representation of Reality in Western Literature.* Translated by Willard R. Trask. Princeton, NJ: Princeton University Press, 1953.

Balthasar, Hans Urs von. *Cosmic Liturgy: The Universe According to Maximus the Confessor.* Translated by Brian E. Daley, S.J. San Francisco: Ignatius Press, 2003.

Barasch, Marc Ian. *Field Notes on the Compassionate Life: A Search for the Soul of Kindness.* Emmaus, PA: Rodale, 2005.

Bassler, Jouette M. *1 Timothy, 2 Timothy, Titus.* Abingdon New Testament Commentaries. Edited by Victor Paul Furnish. Nashville: Abingdon, 1996.

Dominique Barthélemy, O.P. *God and His Image: An Outline of Biblical Theology*, rev. ed. Translated by Dom Aldhelm Dean, O.S.B. San Francisco: Ignatius, 2004.

Bauckham, Richard. *The Bible and Ecology: Rediscovering the Community of Creation.* Sarum Theological Lectures. Waco, TX: Baylor University Press, 2010.

Bauckham, Richard. *Jesus and the Eyewitnesses: The Gospels as Eyewitness Testimony.* Grand Rapids, MI: William B. Eerdmans, 2006.

Bauckham, Richard. "Reading Scripture as a Coherent Story." In *The Art of Reading Scripture.* Edited by Ellen F. Davis and Richard B. Hays. Grand Rapids, MI: William B. Eerdmans, 2003.

Becker, Adam H., and Annette Yoshiki Reed, eds. *The Ways That Never Parted: Jews and Christians in Late Antiquity and the Early Middle Ages.* Minneapolis: Fortress, 2007.

Berry, Thomas. *The Christian Future and the Fate of the Earth.* Maryknoll, NY: Orbis, 2009.

Berry, Thomas. *The Dream of the Earth.* San Francisco: Sierra Club Books, 1988.

Berry, Thomas. *Selected Writings on the Earth Community.* Maryknoll, NY: Orbis, 2014.

Blowers, Paul M. "Unfinished Creative Business: Maximus the Confessor, Evolutionary Theodicy, and Human Stewardship in Creation." In *On Earth as It Is in Heaven:*

*Cultivating a Contemporary Theology of Creation*. Edited by David Vincent Meconi. Grand Rapids, MI: William B. Eerdmans, 2016.

Boomershine, Thomas E. *The Messiah of Peace: A Performance-Criticism Commentary on Mark's Passion-Resurrection Narrative*. Biblical Performance Criticism. Edited by Holly E. Hearon, David Rhoads, and Kelly R. Iverson. Vol. 12. Eugene, OR: Cascade Books, 2015.

Boomershine, Thomas E. "Mark 16:8 and the Apostolic Commission." *Journal of Biblical Literature* 100 (1981): 225–239.

Boyarin, Daniel. *Border-Lines: The Partition of Judaeo-Christianity*. Philadelphia: University of Pennsylvania, 2004.

Boyarin, Daniel. "Semantic Differences; or, 'Judaism'/'Christianity.'" In *The Ways That Never Parted: Jews and Christians in Late Antiquity and the Early Middle Ages*. Edited by Adam H. Becker and Annette Yoshiki Reed. Minneapolis: Fortress, 2007.

Brown, Raymond E. *An Introduction to the New Testament*. The Anchor Bible Reference Library. Edited by David Noel Freedman. New York: Doubleday, 1997.

Calef, Susan A. "Thecla, Acts of Paul And." In *The New Interpreter's Dictionary of the Bible S–Z*. Edited by Katharine Doob Sakenfeld et al. Vol. 5. Nashville: Abingdon, 2009.

Cannato, Judy. *Field of Compassion: How the New Cosmology Is Transforming Spiritual Life*. Notre Dame, IN: Sorin Books, 2010.

Cannato, Judy. *Radical Amazement: Contemplative Lessons from Black Holes, Supernovas, and Other Wonders of the Universe*. Notre Dame, IN: Sorin Books, 2006.

Cavanaugh, William T. *Field Hospital: The Church's Engagement with a Wounded World*. Grand Rapids, MI: William B. Eerdmans, 2016.

Chilton, Bruce. *Rabbi Paul: An Intellectual Biography*. New York: Doubleday, 2004.

Coleman, Richard J. *Eden's Garden: Rethinking Sin and Evil in an Era of Scientific Promise*. New York: Rowman & Littlefield, 2007.

Collins, Adela Yarbro. *Mark*. Hermeneia. Edited by Harold W. Attridge. Minneapolis: Augsburg Fortress, 2007.

Collins, John J. *The Scepter and the Star: The Messiahs of the Dead Sea Scrolls and Other Ancient Literature*. The Anchor Bible Reference Library. Edited by David Noel Freedman. New York: Doubleday, 1995.

Davis, Ellen F. *Proverbs, Ecclesiastes, Song of Songs*. Westminster Bible Companion. Edited by David L. Bartlett and Patrick D. Miller. Louisville, KY: Westminster John Knox, 2000.

Day, Dorothy. *All the Way to Heaven: The Selected Letters of Dorothy Day*. Edited by Robert Ellsberg. New York: Image, 2010.

Day, Dorothy. *The Duty of Delight: The Diaries of Dorothy Day*. Edited by Robert Ellsberg. Milwaukee: Marquette University Press, 2008.

Day, Dorothy. *The Long Loneliness*. New York: Harper & Row, 1952.

Day, Dorothy. "Our Country Passes from Undeclared War to Declared War; We Continue Our Christian Pacifist Stand." *Catholic Worker*, January 1942, www.catholicworker.org/dorothyday/articles/868.html.

Delio, Ilia. *Christ in Evolution*. Maryknoll, NY: Orbis, 2008.

Delio, Ilia. *The Unbearable Wholeness of Being: God, Evolution, and the Power of Love*. Maryknoll, NY: Orbis, 2013.

Delio, Ilia, editor. *From Teilhard to Omega: Co-Creating an Unfinished Universe*. Maryknoll, NY: Orbis, 2014.

Desmond, William. *God and the Between*. Oxford: Blackwell, 2008.

Desmond, William. *The Intimate Universal: The Hidden Porosity among Religion, Art, Philosophy, and Politics*. Insurrections: Critical Studies in Religion, Politics, and Culture. Edited by Slavoj Žižek et al. New York: Columbia University Press, 2016.

Dodd, C. H. *The Parables of the Kingdom*, rev. ed. New York: Charles Scribner's Sons, 1961.

Dowd, Sharyn, and Elizabeth Struthers Malbon. "The Significance of Jesus' Death in Mark: Narrative Context and Authorial Audience." *Journal of Biblical Literature* 125, 2 (2006): 271–297.

Dowling, Robert M. *Critical Companion to Eugene O'Neill: A Literary Reference to His Life and Work*, New York: Facts on File, 2009.

Dunn, James D. G. *The First and Second Letters to Timothy and the Letter to Titus: Introduction, Commentary, and Reflections*. In *The New Interpreter's Bible: A Commentary in Twelve Volumes*. Edited by Leander E. Keck et al., Vol. XI. Nashville: Abingdon, 2000.

Dunn, James D. G. "Form Criticism." In *Searching for Meaning: An Introduction to Interpreting the New Testament*. Edited by Paula Gooder. Louisville, KY: Westminster John Knox, 2009.

Edwards, Denis. "Teilhard's Vision as Agenda for Rahner's Christology." In *From Teilhard to Omega: Co-Creating an Unfinished Universe*. Edited by Ilia Delio. Maryknoll, NY: Orbis, 2014.

Efird, James M. *End-Times: Rapture, Antichrist, Millennium*. Nashville: Abingdon, 1986.

Eisenbaum, Pamela. *Paul Was Not a Christian: The Original Message of a Misunderstood Apostle*. New York: HarperOne, 2009.

Elphinstone, Andrew. *Freedom, Suffering and Love*. London: SCM, 1976.

Fromm, Erich. *The Anatomy of Human Destructiveness*. New York: Holt, Rinehart and Winston, 1973.

Gebara, Ivone. *Longing for Running Water: Ecofeminism and Liberation*. Translated by David Molineaux. Minneapolis: Augsburg, 1999.

Gebara, Ivone. *Out of the Depths: Women's Experience of Evil and Salvation*. Translated by Ann Patrick Ware. Minneapolis: Fortress, 2002.

Fitch, Bob. *The Bob Fitch Photography Archive: Movement for Change*. Stanford University Libraries, https://exhibits.stanford.edu/fitch. Stanford, California, 2018.

Gleick, James. *Chaos: Making a New Science*. New York: Penguin, 1987.

Gooder, Paula. *Searching for Meaning: An Introduction to Interpreting the New Testament*. London: SPCK, 2008.

Gould, Stephen Jay. "A Time of Gifts." *New York Times*, September 26, 2001.

Gould, Stephen Jay. *Hen's Teeth and Horse's Toes: Further Reflections in Natural History*. New York: W. W. Norton, 2010.

Grim, John, and Mary Evelyn Tucker. "Teilhard de Chardin: Biography." American Teilhard Association website, http://teilharddechardin.org/index.php.

Hart, David Bentley. *The Beauty of the Infinite: The Aesthetics of Christian Truth*. Grand Rapids, MI: William B. Eerdmans, 2003.

Hart, David Bentley. *The Doors of the Sea: Where Was God in the Tsunami?* Grand Rapids, MI: William B. Eerdmans, 2005.

Harvey, Nicholas Peter. *Morals and the Meaning of Jesus: Reflections on the Hard Sayings*. Cleveland: Pilgrim Press, 1993.

Hays, Richard B. *The Conversion of the Imagination: Paul as Interpreter of Israel's Scripture*. Grand Rapids, MI: William B. Eerdmans, 2005.

Hennessy, Kate. *Dorothy Day: The World Will Be Saved by Beauty: An Intimate Portrait of My Grandmother*. New York: Scribner, 2017.

Heschel, Abraham Joshua. *The Prophets*. New York: Harper & Row, 1962.

Heschel, Susannah. "Following in My Father's Footsteps." *Vox of Dartmouth*, April 4, 2005.

Heschel, Susannah. *The Aryan Jesus: Christian Theologians and the Bible in Nazi Germany*. Princeton, NJ: Princeton University Press, 2008.

Heschel, Susannah. "What Does It Mean to Be Religious in an Age of Neoliberalism?" *Faith and Leadership*, November 3, 2014.

Hill, Shannon. "Dorothy Day: A Sinner and Nearly a Saint." *Publishers Weekly*, January 23, 2017.

Horsley, Richard A. *Hearing the Whole Story: The Politics of Plot in Mark's Gospel*. Louisville, KY: Westminster John Knox, 2001.

Hunter, David G. "A Decade of Research on Early Christians and Military Service." *Religious Studies Review* 18, 2 (1992): 87–94.

Hurtado, Larry W. *Lord Jesus Christ: Devotion to Jesus in Earliest Christianity*. Grand Rapids, MI: William B. Eerdmans, 2003.

Jewett, Robert. *Paul the Apostle to America: Cultural Trends & Pauline Scholarship*. Louisville, KY: Westminster John Knox, 1994.

Jewett, Robert. *Romans: A Commentary*. Hermeneia: A Critical and Historical Commentary on the Bible. Edited by Eldon Jay Epp. Minneapolis: Fortress, 2007.

Johnson, Elizabeth A. *Women, Earth, and Creator Spirit*. The Madeleva Lecture in Spirituality. Mahwah, NJ: Paulist, 1993.

Johnson, Luke Timothy. *Letters to Paul's Delegates: 1 Timothy, 2 Timothy, Titus*. Valley Forge, PA: Trinity Press International, 1996.

Kasper, Walter. *Mercy: The Essence of the Gospel and the Key to Christian Life*. Translated by William Madges. New York: Paulist Press, 2014.

King, Ursula. *Pierre Teilhard de Chardin: Writings Selected and Introduced*. Modern Spiritual Masters Series. Edited by Robert Ellsberg. Maryknoll, NY: Orbis, 1999.

Lampe, Peter. *From Paul to Valentinus: Christians at Rome in the First Two Centuries*. Translated by Michael Steinhauser. Minneapolis: Fortress, 2003.

Lee, Margaret Ellen, and Bernard Brandon Scott. *Sound Mapping the New Testament*. Salem, OR: Polebridge Press, 2009.

Leithart, Peter J. *Defending Constantine: The Twilight of an Empire and the Dawn of Christendom*. Downers Grove, PA: IVP Academic, 2010.

Leithart, Peter J. *The Four: A Survey of the Gospels*. Moscow, ID: Canonpress, 2010.

Lerner, Michael. *Jewish Renewal: Path to Healing and Transformation*. New York: HarperPerennial, 1994.

Lindsey, Hal. *The Late Great Planet Earth*. Grand Rapids, MI: Zondervan, 1970.

Lohfink, Gerhard. *Is This All There Is? On Resurrection and Eternal Life*. Translated by Linda M. Maloney. Collegeville, MN: Liturgical Press Academic, 2017.

Lohfink, Gerhard. *Jesus of Nazareth: What He Wanted, Who He Was*. Translated by Linda M. Malony. Collegeville, MN: Liturgical Press, 2012.

Luz, Ulrich. *Matthew 1–7: A Commentary*. Translated by James E. Crouch. Hermeneia. Edited by Helmut Koester et al. Minneapolis: Augsburg, 2007.

Malbon, Elizabeth Struthers. *Mark's Jesus: Characterization as Narrative Christology*. Waco, TX: Baylor University, 2009.

MacDonald, Dennis Ronald. *The Legend and the Apostle: The Battle for Paul in Story and Canon*. Philadelphia: Westminster, 1983.

Marcus, Joel. *Mark 1–8: A New Translation with Introduction and Commentary*. The Anchor Bible, Vol. 27. Edited by William Foxwell Albright and David Noel Freedman. New Haven, CT: Yale University Press, 2000.

Marcus, Joel. *Mark 8–16: A New Translation with Introduction and Commentary*. The Anchor Bible, Vol. 27A. Edited by John J. Collins. New Haven, CT: Yale University Press, 2009.

McLaren, Brian D. *We Make the Road by Walking: A Year-Long Quest for Spiritual Formation, Reorientation, and Activation*. New York: Jericho Books, 2014.

Meconi, David Vincent. "Establishing an I-Thou Relationship between Creator and Creature." In *On Earth as It Is in Heaven: Cultivating a Contemporary Theology of Creation*. Edited by David Vincent Meconi, S.J. Grand Rapids, MI: William B. Eerdmans, 2016.

Meyers, Carol. "Women in the OT." In *The New Interpreter's Dictionary of the Bible S–Z*, Vol. 5. Edited by Katharine Doob Sakenfeld et al. Nashville: Abingdon, 2009.

Moltmann, Juergen. *The Coming of God: Christian Eschatology* [*Das Kommen Gottes: Christliche Eschatologie*]. Translated by Margaret Kohl. Minneapolis: Fortress, 1996.

Moltmann, Juergen. "Epilogue: Reverence for the Earth—Friendship with the Wild Beasts; Two Footnotes to Ecological Theology." In *Turning to the Heavens and the Earth: Theological Reflections on a Cosmological Conversion*. Edited by Julia Brumbaugh and Natalia Imperatori-Lee. Collegeville, MN: Liturgical Press, 2016.

Moltmann, Juergen. *God in Creation: A New Theology of Creation and the Spirit of God*. Translated by Margaret Kohl. Minneapolis: Fortress Press, 1993.

Moltmann, Juergen. *In the End—the Beginning: The Life of Hope*. Translated by Margaret Kohl. Minneapolis: Fortress Press, 2004.

Moltmann, Juergen. *The Way of Jesus Christ*. Translated by Margaret Kohl. New York: HarperCollins, 1990.

Nguyen, Joseph. *Apatheia in the Christian Tradition: An Ancient Spirituality and Its Contemporary Relevance*. Eugene, OR: Cascade Books, 2018.

Nicklesburg, George W. E., and James C. VanderKam. *1 Enoch 2: A Commentary on the Book of Enoch, Chapters 37–82*. Hermeneia: A Critical and Historical Commentary on the Bible. Minneapolis: Fortress, 2012.

Ollenburger, Ben C. "The Book of Zechariah: Introduction, Commentary, and Reflections." In *The New Interpreter's Bible: A Commentary in Twelve Volumes*, edited by Leander E. Keck et al. Nashville: Abingdon, 1996.

Ozdemir, Ibrahim. "Toward an Understanding of Environmental Ethics from a Qur'anic Perspective." In *Islam and Ecology: A Bestowed Trust*. Edited by Frederick M. Denny, Richard C. Foltz, and Azizan Baharuddin. Religions of the World and Ecology. Cambridge, MA: Harvard University Press, 2003.

Pesch, Rudolf. *Das Markusevangelium II Teil: Kommentar Zu Kap. 8,27–16,20*. Herders Theologischer Kommentar Zum Neuen Testament, Vol. 2/2. Herausgegeben von Rudolf Schnackenburg, Alfred Wikenhauser. Freiburg: Herder, 1977.

Prothero, Stephen. *American Jesus: How the Son of God Became a National Icon*. New York: Farrar, Straus and Giroux, 2003.

Rahner, Karl. "106. Women and the Priesthood," In *The Content of Faith: The Best of Karl Rahner's Theological Writings*. Translated by Harvey D. Egan, S.J. Edited by Karl Lehmann and Albert Raffelt. New York: Crossroad, 1992.

Rhoads, David. *Israel in Revolution: 6–74 C.E.* Philadelphia: Fortress, 1976.

Rhoads, David. *Reading Mark: Engaging the Gospel*. Minneapolis: Fortress, 2004.

Rhoads, David, Johanna Dewey, and Donald Michie. *Mark as Story: An Introduction to the Narrative of a Gospel*. Third edition. Minneapolis: Fortress, 2012.

Rohr, Richard. *Simplicity: The Freedom of Letting Go*. Translated by Andreas Ebert. Revised English-language edition. New York: Crossroad, 2003.

Rossing, Barbara R. *The Rapture Exposed: The Message of Hope in the Book of Revelation*. New York: Basic Books, 2004.

Rovelli, Carlo. *Seven Lessons on Physics*. Translated by Simon Carnell and Erica Segre. New York: Riverhead, 2014.

Russell, Heidi Ann. *Quantum Shift: Theological and Pastoral Implications of Contemporary Developments in Science*. Collegeville, MN: Liturgical Press, 2015.

Sacred Congregation for the Doctrine of the Faith. "Declaration on the Question of the Admission of Women to the Ministerial Priesthood." Rome, 1976.

Schaff, Philip. *History of the Christian Church*. 12th edition. Vol. II: Ante-Nicene Christianity. A.D. 100–325. New York: Charles Scribner's Sons, 1914.

Schmidt, Karl Ludwig. *Die Rahmen Der Geschichte Jesu: Literarkritische Untersuchungen Zur Aeltesten Jesusueberlieferung*. Berlin: Trowitzsch & Sohn, 1919.

Schnackenburg, Rudolf. *Das Evangelium Nach Markus*. Geistliche Schriftlesung: Erlaeuterungen Zum Neuen Testament Fuer Die Geistliche Lesung. Herausgegeben von Karl Hermann Schelkle, Wolfgang Trilling, Heinz Schuermann. Vol. 2/2. Dusseldorf: Patmos, 1971.

Slattery, John P. "Dangerous Tendencies of Cosmic Theology: The Untold Legacy of Teilhard de Chardin." *Philosophy and Theology* 29, 1 (2018): 69–82.

Spadaro, Antonio, S.J. "A Big Heart Open to God: An Interview with Pope Francis." *America: The Jesuit Review*, September 30, 2013.

Stark, Rodney. *The Rise of Christianity: How the Obscure, Marginal Jesus Movement Became the Dominant Religious Force in the Western World in a Few Centuries*. San Francisco: HarperCollins, 1997.

Stump, Eleonore. *Wandering in Darkness: Narrative and the Problem of Suffering*. Oxford: Clarendon, 2010.

Suchocki, Marjorie Hewitt. *God Christ Church: A Practical Guide to Process Theology*. New Revised Edition. New York: Crossroad, 1992.

Talbert, Charles H. *Reading Corinthians: A Literary and Theological Commentary on 1 and 2 Corinthians*. New York: Crossroad, 1992.

Teilhard de Chardin, Pierre. *The Making of a Mind: Letters from a Soldier-Priest 1914–1918*. Translated by Rene Hague. New York: Harper & Row, 1965.

Teilhard de Chardin, Pierre. *Letter to Christophe de Gaudefroy, 7 October 1929, Lettres inédites,* 80. American Teilhard Association, http://teilharddechardin.org/index.php/teilhards-quotes.

Tanenbaum, Marc H. "Heschel and Vatican II: A Memorial Symposium in Honor of Rabbi Abraham Joshua Heschel." *Archives of the American Jewish Committee*. New York, February 21, 1983.

Thompson, Virgil. *The New Testament in Contemporary Life*. Dubuque, IA: Kendall Hunt, 2011.

Tkacz, Catherine Brown. *Αληθεια Ελληνικη: The Authority of the Greek Old Testament*. Etna, CA: Center for Traditionalist Orthodox Studies, 2001.

Tolbert, Mary Ann. *Sowing the Gospel: Mark's World in Literary and Historical Perspective*. Minneapolis: Fortress, 1989.

Tollefsen, Torstein T. "Christocentric Cosmology." In *The Oxford Handbook of Maximus the Confessor*. Edited by Pauline Allen and Bronwen Neil. Oxford: Oxford University Press, 2015.

Torjesen, Karen Jo. "Reconstruction of Women's Early Christian History." In *Searching the Scriptures: One: A Feminist Introduction*. Edited by Elizabeth Schuessler Fiorenza. New York: Crossroad, 1993.

Trobisch, David. *Paul's Letter Collection: Tracing the Origins*. Minneapolis: Fortress, 1994.

Verhey, Allen. *Remembering Jesus: Christian Community, Scripture, and the Moral Life*. Grand Rapids, MI: Eerdmans, 2002.

Waetjen, Herman. *The Origin and Destiny of Humanness: An Interpretation of the Gospel According to Matthew*. Corte Madera, CA: Omega, 1976.

Wink, Walter. *Engaging the Powers: Discernment and Resistance in a World of Domination*. Minneapolis: Fortress, 1992.

Wogaman, J. Philip. *Christian Ethics: A Historical Introduction*. Louisville, KY: Westminster John Knox, 1993.

*Women's Bible Commentary*. Expanded Edition with Apocrypha. Edited by Carol A. Newsom and Sharon H. Ringe. Louisville, KY: Westminster John Knox, 1998.

Wordelman, Amy L. "Everyday Life: Women in the Period of the New Testament." In *Women's Bible Commentary*. Edited by Carol A. Newsom and Sharon H. Ringe. Louisville, KY: Westminster John Knox, 1998.

Wright, N. T. *How God Became King*. New York: HarperOne, 2012.

Wright, N. T. *Jesus and the Victory of God*. Christian Origins and the Question of God, Vol. 2. Minneapolis: Fortress, 1996.

Wright, N. T. "Narrative Theology: The Evangelists' Use of the Old Testament as an Implicit Overarching Narrative." In *Biblical Interpretation and Method: Essays in Honour of John Barton*, edited by Katharine J. Dell and Paul M. Joyce. Oxford: Oxford University Press, 2013.

Wright, N. T. *The New Testament and the People of God*. Christian Origins and the Question of God, Vol. 1. Minneapolis: Fortress, 1992.

Wright, N. T. *Paul and the Faithfulness of God*. Christian Origins and the Question of God, Vol. 4. Minneapolis: Fortress, 2013.

Wright, N. T. *The Resurrection of the Son of God*. Christian Origins and the Question of God, Vol. 3. Minneapolis: Fortress, 2003.

Wright, N. T. *Surprised by Hope: Rethinking Heaven, the Resurrection, and the Mission of the Church*. New York: HarperOne, 2008.

Zagano, Phyllis. *Holy Saturday: An Argument for the Restoration of the Female Diaconate in the Catholic Church*. New York: Crossroad, 2000.

Zagano, Phyllis. "A Woman on the Altar: Can the Church Ordain Women Deacons?" *U.S. Catholic*, November, 2011.

# Index